RELATIONS BETWEEN LOGIC and MATHEMATICS in the WORK of BENJAMIN and CHARLES S. PEIRCE

ALISON WALSH

Docent
Press

DOCENT PRESS
Boston, Massachusetts, USA
www.docentpress.com

Docent Press publishes monographs and translations in the history of mathematics for thoughtful reading by professionals, amateurs and the public.

Cover design by Brenda Riddell, Graphic Details, Portsmouth, New Hampshire.

Contents

List of Figures

List of Tables

CHAPTER 1

Introduction

1.1. Philosopher and Logician

As both philosopher and logician, Charles Sanders Peirce (1839–1914) was neglected by the philosophical community of his time and misunderstood by the logicians. He was one of the principal founders of modern logic and the inventor of the influential philosophy of pragmatism. Although his influence is currently being recognized by modern logicians, with the publication of a volume of essays and papers[1] arising out of the important Sesquicentennial International Congress held at Harvard University in 1989 to celebrate his birth in 1839, Peirce is still the victim of historical ignorance. The pioneering algebraic logic of Peirce and Schröder was largely eclipsed until the 1940s by the mathematical logic of Gottlob Frege, Guiseppe Peano and Bertrand Russell, with the result that we are only now coming to realize and discover the power of Peirce's logical work. In this work we start with his father Benjamin Peirce's linear associative algebra and then consider this and other early influences on the logic of Peirce. A discussion of the early algebraic logicians such as Boole, Jevons and De Morgan follows, culminating in a detailed analysis of Peirce's seminal paper "Description of a Notation for the Logic of Relatives." His further developments of the 1880s, including quantificational logic are also traced. At the end of his life, Peirce looked to his graphical logic system - the existential graphs - to provide the logic of the future.

---

[1] See [98].

1

Even though commentators have recognized Peirce as one of the foremost American logicians, the mathematical techniques that he used have not been closely considered. I provide such an analysis, in particular looking at the problem-solving techniques employed not only by Peirce but also by his graduate students Christine Ladd-Franklin, and Oscar Howard Mitchell and his logical successor Ernst Schröder. The notations and philosophy of these logicians have been previously documented, but any study of the algebraic methods used by these logicians when they came to apply their logics has until now, been lacking.

The main publications of Charles Peirce's work on logic that have been cited in this thesis are *Writings of Charles S. Peirce, A Chronological Edition* edited by Max Fisch, Matthew Moore[2], and *Collected Papers of Charles Sanders Peirce*, Volumes 1–6 edited by Charles Hartshorne and Paul Weiss,[3] Volumes 7–8 edited by Arthur Burks[4]. Peirce's unpublished papers are referenced by MS for manuscript followed by the folder number assigned in [**172**]. Victor Lenzen brought the Peirce manuscripts to Harvard from Peirce's home after his death in 1914. The Department of Philosophy at Harvard then arranged the preparation of the *Collected Papers* that were published in six volumes between 1931 and 1935.

However, this was a poor edition which had an adverse as well as a positive effect on Peirce scholarship with editors Paul Weiss and Charles Hartshorne deleting or changing sections of the manuscripts. This section highlights the fact that Peirce's own signs which were carefully chosen as icons for the sixteen binary connectives were replaced by more conventional symbols of the editors' choosing. Not only were they the wrong signs but they lacked any attempt at iconicity. Arthur Burks produced volumes 7 and 8 in the mid-1950s and Max Fisch directed the sorting of the papers into proper order for the first time in 1972, starting the Peirce Edition Project which plans to publish thirty volumes of the *Chronological Edition* of his writings. It is this latter edition which has led to a fruitful study of Peirce's work in logic. Some of Peirce's pioneering studies are inevitably ambiguous here and there and I have chosen reasonable interpretations in certain places, particularly in the 1870 paper "Description of a Notation for the Logic of Relatives". There now follows a brief summary of the story.

---

[2][**55**], [**56**], [**57**], [**58**], [**59**], [**60**]
[3][**82**], [**83**], [**84**], [**85**], [**86**], [**87**]
[4][**24**], [**23**]

1.2. Benjamin Peirce's *Linear Associative Algebra*

Benjamin Peirce's *Linear Associative Algebra* first published in lithographic form in 1870,[5] was his main algebraic work and is important because it marks the first stage in the development of modern day linear algebra. The main points of *Linear Associative Algebra* are outlined in Chapter 2, along with some biographical comments about this remarkable man. In painstaking detail, after a few pages of definitions and axioms, Benjamin listed all possible linear associative algebras in the form of their multiplication tables for systems of up to six units resulting in an almost overwhelming definition of 163 algebras and six subcases.

Another feature of *Linear Associative Algebra* is the extreme brevity of its proofs. Hubert Newton testifies: "...[his] demonstrations are given only in outline being in respect of fullness the entire opposite of Euclid" [**130**, p. 168]. This eccentric style as shown in the the multiple expositions of the algebras of *Linear Associative Algebra* was a distinct disadvantage. As his student and then colleague Thomas Hill affirmed, Benjamin had a "...habit of using simple conceptions, axioms and forms of expression, without reference to established usage to produce demonstrations ...[of] exceeding brevity"[6]. This family trait was handed down to Charles.

In Chapter 2, we supply an analysis of two such proofs in particular the proof of the axiom that "In every linear associative algebra there is at least one idempotent or one nilpotent expression." We also analyze many of the multiplication tables for the algebras supplying any calculations omitted by Benjamin and pointing out errors that have never before been corrected. This is a valuable exercise in both understanding the reasoning behind Benjamin Peirce's 169 multiplication tables where the algebraic explanations are often omitted, and in highlighting errors in his own working both in the 1879 lithograph and in the 1881 *American Journal of Mathematics* version.

1.3. Later Papers: the Development of Peirce's Algebraic Logic

Algebraic logic as developed in different ways by George Boole (1815–1864) and Augustus De Morgan (1806–1871) attempted to express the laws of thought or the

---

[5]It was later edited by C. S. Peirce and reprinted in *American Journal of Mathematics* 1881, v. 4, p. 97–229.

[6]See [**168**, p. 144].

processes of thinking and logical deduction in the form of mathematical equations. Using the traditional syllogism as introduced by Aristotle and developed by the medieval scholars as a starting point, these logicians were interested in problem solving and deducing conclusions by applying mathematical techniques. In contrast to mathematical logic, Boole used letters to represent classes of objects, rather than sets of objects. There was no elementhood relation but only the relations of proper and improper inclusion were used.

Charles Peirce's primary interest in algebraic logic came from the logic of George Boole. In Chapter 3, we highlight the three main areas where Boole's logic departed from an arithmetic system, namely the operation of division, the index law and the interpretation of symbols.[7] By analyzing Peirce's early Harvard Lectures in terms of the definitions of the logical terms, the operations, zero and unity, and comparing them with Boole's definitions in both *Mathematical Analysis of Logic* of 1847, and *Laws of Thought* of 1854, we show that Peirce seems to be working from *Laws of Thought*, rather than the earlier work. We also analyze in detail an example of logical elimination in Peirce's "Harvard Lecture III" of 1865, using Boole's Development Theorem, any proof being completely omitted by Peirce. The example given is that of the well-known syllogism

> All men are animals.
> Socrates is a man.
> Therefore Socrates is an animal.

In contrast to his earlier complete acceptance of Boole's logic, Peirce now began to improve its deficiencies – "enormous deficiencies" as he was to say – in "Harvard Lecture VI" of 1865, which soon became apparent. He was to extend Boole's calculus by providing the missing operation of division, for which we suggest the definition

$$x = b/a = b + v(1 - a),$$

where v refers to the indeterminate class meaning SOME, ALL OR NONE, thus forming Boole's main method of quantification. We analyze the meaning of v in Boole's logic showing that Boole had different interpretations for v at different times. We also show that Peirce made a serious misreading of Boole's view of the nature of numerical coefficients which resulted in an error when he came to demonstrate a Boolean example.

---

[7]Boole himself claimed there was only one point of divergence in the laws of logic and those of number. He probably had in mind the Index Law i.e. $x^2 = x$ in [**15**, p.11].

It is a matter of some interest that this error involved an equation of the form $x + x = x$. Boole expressly ruled out such equations but Peirce was to incorporate the rule in his new operation of addition, which he discovered independently of but subsequently to W. Stanley Jevons (1835–1882). This operation extended Boolean addition to cover non-disjoint classes.

In Chapter 3, we outline the logic of Jevons and consider his approach to the relationship between mathematics and logic, which was completely different from that of Boole's. In fact in his *Formal Logic* of 1864, Jevons was one of the first proponents of the logistic view that logic is the basis of mathematics. However his method of inference which consisted of taking all possible combinations of the terms of the premises and their complements, combining each of these terms separately with both sides of a premise and then eliminating to find the solution, is a longer and more tedious method than Boole's methods. It is interesting to note that Jevons disagreed with Boole in two key areas.

Firstly, he proposed the operation of addition between classes, which can be defined as $a + b = a + b - ab$ for non-distinct classes a and b. This culminated in Jevons' law of unity $a + a = a$, which Boole completely rejected. Secondly, because of the above law of unity, Jevons objected to both addition and subtraction as processes of logic, and although he used the operations of multiplication and addition in his own algebraic logic, he preferred to consider them as the logical operations of combination and separation.

So by 1867, when Peirce came to write "On an Improvement in Boole's Calculus" he was aware that a major limitation of Boole's algebraic logic was in the area of quantification, in that it could not properly express particular or categorical propositions such as *Some X is Y*. To this end he defined his operation of addition and its inverse, logical subtraction. Furthermore we show that he introduced a new operation of logical division, probably to complete the algebraic analogy. However shortly after this, he discovered the work of De Morgan on the copula.

1.4. The Theory of Relations

De Morgan had realized the inadequacies of syllogistic logic and claimed that some way of representing relations other than the identity relation was needed. His theory of relations involved expressing inferences in logic in terms of composition of relations:

$$X..LY \qquad\qquad X \text{ is an } L \text{ of } Y$$
$$X.LY \qquad\qquad X \text{ is not an } L \text{ of } Y$$

Peirce developed and extended De Morgan's work on the theory of relations. We look specifically at a series of papers written by De Morgan entitled "On the Syllogism: I" to "On the Syllogism: IV", written in the period 1846–1849. These papers contain his logical development of relations and are of particular significance to Peirce.

De Morgan issued a challenge to contemporary logicians: Deduce *Every head of a man is the head of an animal* from *Every man is an animal* using only the traditional identity copula. Charles Peirce found this irresistible. Peirce combined his own theory of relations with an extended and improved version of Boole's algebraic logic in a seminal paper: "Description of a Notation for the Logic of Relatives, resulting from an Amplification of the Conceptions of Boole's Calculus of Logic" [**149**][8]. For the first time, Boole's part/whole class calculus was combined with De Morgan's theory of relations to form an algebraic logic equivalent to today's predicate logic. This thesis also explains Peirce's mysterious logical differentiation. What for example is the interpretation of

$$d x^2 = 2x^1 d x,$$

where each symbol represents an operation or class? Although the algebraic machinery to support logical differentiation is highly developed, Peirce omitted to provide the verbal translation, which has led to much speculation by current scholars.

1.5. Description of a Notation for the Logic of Relatives

Having incorporated a theory of relations into Boolean algebraic logic, Peirce developed a powerful notation and problem-solving algebraic logic equivalent to first order predicate logic. This was first published in *Logic of Relatives*. A weakness of Boole's logic, soon realized by Peirce, is that it is not able to express satisfactorily categorical or particular propositions of the form SOME *X* IS *Y*. It was precisely this problem of quantification that inspired Peirce to make the Amplification. Sections of *Logic of Relatives* were amended just before being sent to the printers, mainly in the

---

[8]This work will be referred to as *Logic of Relatives* henceforth.

light of De Morgan's superscript notation for quantification, as used in his operation of involution. It is not therefore surprising that Peirce makes so many errors in his proofs. In many cases he cites incorrect formulae numbers which makes it even more difficult to follow his sketchy proofs. We suggest that this was caused by the fact that additional formulae inserted just prior to printing were not taken account of in the final numbering. In Chapter 4, we supply an analysis and expansion of some of his proofs, tracking down the correct formulæ where appropriate.

There follows a discussion of Peirce's logical terms. A term is used by Peirce in an analogous manner to an algebraic term, but while such a term represents a quantity, a logical term represents either a predicate in first order logic or a proposition in second order logic. We clarify the main misconceptions of *Logic of Relatives*, one of which is the confusion surrounding his interpretation of the individual, defined as an absolute term. The question involved here is what is an absolute term and is it a class? Our answer is that an absolute term, e.g., WOMAN is a representative member of a class denoting that class, but it is only an instance or ideal member of that class. So Peirce's absolute term is neither a class nor a specific individual. However such absolute terms may represent specific individuals, as in a linear combination, while a class is made up of an aggregation of specific individuals.

The second main misconception is the blurring of the boundaries of relation and relative term. Note that Peirce's algebraic logic is the logic of relatives not De Morgan's logic of relations. Peirce first defined the relative term s as WHATEVER IS THE SERVANT OF _____[9]. Some commentators have identified these relative terms with relations. Others with classes associated with the relation. We are more inclined to this latter view. However relative terms can be understood as significations of linguistic items which are derived from verb phrases (e.g. serves a master) with a blank for a noun, e.g., WHATEVER IS A SERVANT OF_____. This changed in 1882 when he defined a relative term as a class of ordered pairs, i.e., as what we recognize today as a relation. Any difficulty that arises in the minds of Peirce commentators lies back in the 1870 *Logic of Relatives* paper where he seems to have confused relations with relative terms.

We show clearly in Chapter 4 that the real problem is in fact the quite different issue of the general confusion between absolute terms and relative terms. Peirce used absolute terms and relative terms interchangeably. In order to achieve this he used the comma operator that converts absolute terms to relative terms. For example, consider

---

[9]See [**83**, p. 369].

the absolute term SERVANT, s. This is transformed by the comma operator into "s," or WHATEVER IS THE SERVANT OF_____. Peirce scholars have claimed that he confused the *relative term*, WHATEVER IS THE SERVANT OF _____, with the *class of servants* which represents the relation, so that all relative terms are in fact really relations. My point is that he uses SERVANT as an absolute term and not as a class at all but a representative member of a class. As such, it is not a direct representation of a relation.

Peirce later moved away from relative terms to emphasize relations between classes. The change is most clearly stated in his 1897 paper "The Logic of Relatives" as the change from classes to relations or as he wrote:[10]

> The best treatment of the logic of relatives, as I contend, will dispense altogether with class names and only use... verbs.

It was probably effected because propositions could now be simplified to a variable signifying the relation, subscripts signifying the individuals and a quantifying symbol.

This key area of Peirce's logic, namely exactly what did he mean by a relative term has also been addressed by Burch in [21] which claims that it makes little difference whether we talk of relations or of relatives. Since Peirce was clear in his aim of separating the syntax from the semantics of his logic, it is conceivable that relative terms indicate classes or functions or even objects depending on the universe of discourse taken. Burch agrees that Peirce does focus on the class definition of relative term in at least his 1870 paper, but argues that there is no reason to deny that Peirce is constructing a logic of relations since he is discussing relations by concentrating attention on their bearers. By using a graphical formulation, Burch claims to show that Peirce's logic of relatives as expressed in 1870 is at least as powerful in expressive capability as first order predicate logic with identity.

In the later sections of *Logic of Relatives*, Peirce introduces a new and mysterious process – logical differentiation. Directly analogous to mathematical differentiation, it uses logical terms instead of mathematical variables. Chapter 4 takes an original turn when we introduce new interpretations for these variables that serve to clarify Peirce's process. Associated with and essential to logical differentiation, is an understanding of his use of logical terms, his process of logical multiplication, the logical

---

[10]See [**160**, p. 290].

analogy to the binomial theorem, infinitesimal relatives, the concepts of numerical co-efficients and the number associated with each term. All these concepts are discussed in this section. We also analyze the algebraic development of logical differentiation and consider in depth one application of the process. Peirce provided here an ingenious analogy to mathematical differentiation. We will follow the process and identify some errors made by him in this section. This part of *Logic of Relatives* comes just before that on backward involution, which we know was added quickly to the final proof of the paper. The fact that these sections were written without revision could account for the fact that Peirce did not give English translations. The result was to make the work even more obscure. By providing such interpretations that follow simply from the definitions, we shed some light on this mysterious process.

1.6. The Theory of Quantification

In Chapter 5, we consider Peirce's algebraic logic post *Logic of Relatives*, including a review of his main successors Oscar Mitchell (1851–1889), Christine Ladd-Franklin (1847–1930) and Ernst Schröder (1841–1902). There was a decade of little further algebraic work apart from the introduction of three main innovations, viz., duality, modal logic and transaddition. In 1883, Charles Peirce published a volume entitled *Studies in Logic, by Members of the Johns Hopkins University* [120]. Looking in more detail at the work of Ladd-Franklin and Mitchell who were graduate students taking Peirce's logic course at the Johns Hopkins University, We analyze their different versions of algebraic logic for problem solving. This is an area that has been neglected by historians. Not only do we clarify their algebraic methods but we have also been able to identify errors in working or minor slips in specific logical problems that surprisingly have not been discovered before. We summarize the development of the quantifier and look at the advantages of Peirce's quantification theory over his algebraic logic.

Ladd-Franklin's algebraic logic is then considered. Not only was this singularly lacking in terms of quantification, but there were also no relations or relative terms, only the traditional identity copula. However she did clearly deal with at least existential quantification. Her operation of $\overline{\overline{V}}$ was used almost like a quantifier symbol to denote that a particular predicate or proposition did not exist. Existence was not so well defined, as xV denoted that x is at least sometimes existent. Another point of note is their use of modal values. Truth-values are used for *propositions* such that aVb denotes that propositions a and b have been at some moment of time both true, whereas for *predicates* a and b, aVb denotes that a and b are co-existent.

A discussion of Mitchell's innovative logical ideas follows. One of these was to separate the universe of class terms and the universe of relative terms thus showing the way to a multi-dimensional logic. By using different universes of discourse for predicates and propositions, Mitchell overcame a major failing of Boolean algebra in its difficulty in expressing mixed hypothetical and categorical statements. Another aspect of Mitchell's work that is considered is his use of indices to represent quantification because this gave Peirce the key to his own quantificational logic. Mitchell did not use the symbols Π and Σ for his quantifiers. These were used, as Peirce had used them previously in *Logic of Relatives*, to denote infinite sums and products in linear combinations of logical terms, not as quantifiers.

The last part of Chapter 5 comprises a discussion of some aspects of the work of Ernst Schröder. By 1879 Schröder was familiar with Peirce's logic of relatives and incorporated this in his later logical work. He was later influenced by Peirce's theory of quantification and so was able to progress from the predicate logic of Boole to the algebraic logic of Peirce in terms of incorporating firstly his logic of relations and secondly his quantificational logic. By this wholesale adoption of Peircean logic, he proved himself to be a true logical successor. However the influence seems to have been mainly one way (apart from features such as duality) which is shown in Peirce's rather arrogant dismissal of his logic, outlined in correspondence and other writings as detailed in this section.

Having said that, we also note the main differences between the logical views of Peirce and Schröder. Peirce's criticism of Schröder was over the Hypothetical-Categorical debate. Schröder held that all hypothetical propositions of the form IF *A* THEN *B* can be reduced to categoricals ALL *A* IS *B*, but not vice versa. Peirce held that categoricals are modifications of the form of hypotheticals. This disagreement arose because of the fundamentally different aims of the two logicians. Peirce was interested in a calculus of logic that covered individuals, classes and propositional logic, whilst Schröder wished to differentiate between them, e.g., he favored separate symbols for classes and for propositions. This section also considers a main area of neglect by historical logicians, Schröder's problem-solving techniques. We are able to supply an analysis of his algebraic methods used for such a purpose. One logical problem analyzed has been previously considered when discussing the problem-solving of Ladd-Franklin who had the same problem but used completely different algebraic methods and notation.

In the final chapter, we outline the influences of Benjamin Peirce, Boole, De Morgan and Mitchell on Charles Peirce's algebraic logic. We consider his main achievements and trace the new path taken by Peirce in developing post 1897, a graphical form of logic called the existential graphs. This was predicted by Peirce to be the logic of the future and he did very little algebraic logic work after this date. These graphs were not at all algebraic but were a form of logical diagram. Initially inspired by the amateur British mathematician and philosopher Alfred Bray Kempe (1849–1922), Peirce moved from his early form, the entitative graphs, to his final form, the existential graphs. These diagrams represented to Peirce a way of analyzing logical inference by a method superior to his previous algebraic systems. Finally, we note the influence of Peirce on later logicians.

1.7. Literature Review

Much has been written about Charles Peirce, the philosopher, but little on the algebraic methods used in his logic. Contemporary historians of logic have until recently ignored or downplayed the value of the algebraic logic tradition of the nineteenth century, partly because it was heavily eclipsed by the mathematical logic of Russell, Zermelo and others. In the anthology *From Frege to Gödel* [92], intended as a representative documentary history of the formative years of mathematical logic, the algebraic tradition is virtually ignored and deliberately so. Historical surveys devote very little attention to the algebraic tradition.

For example, Bochenski's history of logic [13] devotes only some ten pages to the Boolean calculus and some twelve pages to the logic of relations, most of which focus on Russell's work rather than that of De Morgan, Peirce, and Schröder, while the historical survey of Kneale [109] devotes all of thirty pages to Boolean algebra and the logic of relations. Peirce's logic has also suffered because until the Peirce Project Edition series of his published works, *The Chronological Writings*, appeared, the only published Peirce material available was that of the 1930s edition, *The Collected Papers*. This edition is an inferior version with many omissions and unnecessary interpolations from the editors. To some extent this omission is being rectified by [74] which has sections on Peirce, Schröder and the main proponents of algebraic logic.

The main texts used for a general survey of Peirce's algebraic logic have been [129] for the development of Peirce's philosophy and logic, [109] and [115] for a historical review. However, these latter works are mainly a reformulation of Peirce's logic in terms of set-theoretical notation with no study of important results or any consideration of the problem-solving techniques used. [131], [194] and [67] were also used as general studies of the mathematical and logical climate of the period. Brent book, *Charles Sanders Peirce, A Life* [17], was used to provide biographical details of Peirce's tragic life.

Regarding Benjamin Peirce's *Linear Associative Algebra*, [19] shows how Charles Peirce's relative multiplication, his central mode of combination of concepts was derived from the multiplication schema for the linear associative algebras developed by his father. [168] and [132] support the case that Benjamin built upon and extended the work of his British contemporaries and adopted their symbolical approach to algebra and justified his work through his strong religious conviction. These papers also suggest reasons for the poor reception of *Linear Associative Algebra*. This latter point is also picked up by [73] which throws new light on its preparation and publication. The main critiques of *Linear Associative Algebra* are [88] and [195] which are attempts to reformulate *Linear Associative Algebra* and extend its results in terms of hypercomplex numbers using matrix theory, [88] and [189] also review the work of contemporary mathematicians in the same field.

Charles Peirce's early works are examined in [126] which traces the influence of Boole's algebra on his developments in logic. Heath's edition of De Morgan's series of articles on the Syllogism was extensively used [90]. [124] concentrates on De Morgan's theory of relations whilst [122] concentrates on Peirce's own development of the logic of relations, looking for the influences of De Morgan within this development. [121] and [18] were also used in this context. As far as the history of quantification theory is concerned we have to mention [47] which illuminates the contribution of Oscar Mitchell's pioneering work on the introduction of the quantifier. Recent works on this topic include [16] which summarizes the transition from Peirce's algebraic logic to his quantificational logic and [125] which analyzes the quantificational logic in terms of its power and expressibility.

Invaluable aids to understanding the relationship between Peirce and Schröder, their similarities and differences, proved to be [43] and [97] which details the Peirce-Schröder correspondence. Any analysis of Schröder's logic is largely missing, again to be addressed by [74], but this thesis does supply an examination of his problem-solving techniques that were applied to his notation and logical terms thus taking a small step towards clarifying his logic. [136] was used to examine the trends and influences in Schröder's logic and to identify the fact that his algebra and logic of relatives became the pasigraphic key for the creation of a scientific universal language. [46] was used to supplement the bibliographical details of Schröder's life and work.

The graphical logic of Peirce is extensively covered by two texts, [171] and [205], both of which consider the logical diagrammatic systems which consist of alpha, beta and gamma graphs. [171] in particular, which is in book form, is now raising the awareness of the logical and mathematical communities to this topological turn. In terms of current research, the recently published work *Studies in the Logic of Charles Sanders Peirce* edited by Houser, Roberts and Van Evra should be noted. This collection of essays is a result of the Charles S. Peirce Sesquicentennial International Congress held at Harvard University in 1989 which brought together over 450 scholars from twenty-six countries to commemorate Peirce's birth on 10 September 1839.

The continuing importance and usefulness of his ideas are brought out by this volume. For example we have [21] and [171] which are applications of Peirce's existential graphs. We also have [191] which uses the existential graphs as the foundation for a system of conceptual graphs that provide a logic for representing the semantic structure of natural language. Finally other papers in this volume which have been used as sources to clarify the relation between logic and mathematics of the algebraic logicians and specifically Peirce's own position on this, are [98], [73], [107], and [114].

1.8. Differences between Algebraic Logic and Mathematical Logic

Algebraic logic has been neglected by many historians of logic, largely because of its eclipse by mathematical logic. The main difference in these two traditions is that the algebraic logicians applied algebraic techniques to express and develop logic, whereas the mathematical logicians in varying degrees, held that logic was best expressed using set theoretical concepts and notation in the form of an axiomatic system, as opposed to the part-whole theory of collections that supports algebraic logic. Another belief held by many mathematical logicians was that such a logical system would form the basis for a firm foundation for mathematics. This has been given the term *logicism*. Unfortunately it has been the case that many historians have equated mathematical logic with the logicist programme. This is a false assumption. Schröder although working in the algebraic logic tradition held logicist views. Peano too is a counterexample in the mathematical logic camp, as he was not a logicist.

The algebraic and logical methods that developed in France after the Revolution concerned semiotic ideas that emphasized the clarity of signs and the use of algebraic techniques to other branches of mathematics and to logic.[11] These algebraic methods influenced De Morgan in his work on functional equations in 1836 and then in his logic of relations in 1847. Boole was also influenced firstly in his work on differential operators of 1844 and then in his algebraic logic in 1847. Peirce and Schröder from 1870 onwards followed this algebraic tradition, and extended it to incorporate a theory of quantification.

In "Peirce Between Logic and Mathematics" Grattan-Guinness traces the strand of mathematical logic that began with a partial reaction against the algebraic methods of Lagrange by Cauchy (1789–1857) who developed mathematical analysis, using a method of limits to embed the calculus, the theory of functions and the convergence of series. Grattan-Guinness writes:[12]

> The main aim was to improve the level of rigour in these subjects, and one aspect is worth noting here: Cauchy greatly improved the logic of specifying necessary and/or sufficient conditions under which theorems were held to be true.

In Germany, Karl Weierstrass in the 1860s adopted and improved on these new methods. Cantor (1845–1918) developed set theory as an extension of mathematical analysis. Peano in the 1880s proved to be the link between mathematical analysis and mathematical logic, formalizing the symbolic notation used. From the 1900s Russell and Whitehead saw a means of basing all mathematics in the set theoretical terminology and axioms that Peano had partly founded, and Frege had developed a theory of quantification prior to that of Peirce in his *Begriffsschrift* of 1879, and also followed

---

[11]See [**68**, pp. 73–74] for a review of the development of logic in France after the Revolution.
[12]See [**73**, p. 27].

the logicist tradition in that he claimed that some parts of mathematics could be based in logic.

The two traditions of algebraic logic and mathematical logic highlight the relationship between logic and mathematics.[13] In algebraic logic, the laws, duality properties and symbols of mathematics were used to develop systems of logic and inference. However, many algebraic logicians with the exception of Schröder considered the disciplines of mathematics and logic to be entirely separate and distinct. They made new developments in logic by applying algebraic principles. Peirce himself, believed that by developing logic in this way, new mathematical methods could be discovered and understood.

---

[13]A full account of the similarities and differences between these traditions can be found in [**73**, pp. 28-32].

CHAPTER 2

Benjamin Peirce's *Linear Associative Algebra*

2.1. Benjamin Peirce – The Man

Benjamin Peirce (pronounced p-IH-r-s), was born on the 4th of April 1809 in Salem, Massachusetts. He came from Puritan stock. The American branch of the Peirce family originated from the descendants of a weaver named John Pers of Norwich, England who emigrated to the United States in 1637. Benjamin was the third child and second son of his father, also called Benjamin. This Benjamin graduated from Harvard College in 1802, served in the Massachusetts State Senate and was, before his death, librarian of Harvard College.

An early contact was Nathaniel Bowditch, then the most important American mathematician, whom Benjamin met through a school friend Henry Bowditch, Nathaniel's son, at the Salem Private Grammar School. This was the decisive factor that led Benjamin to dedicate himself to mathematics. After Benjamin corrected some supposed errors in Bowditch's work, the older mathematician took an interest in the young Benjamin Peirce.[1] The family connections with Harvard continued with Peirce entering Harvard in 1825 and graduating in 1829. For the next two years he taught at George Bancroft's Round Hill School in Northampton before he returned as a tutor in mathematics at Harvard College.

His early mathematical work under the influence of Bowditch dealt chiefly with geometry and with analysis, particularly as applied to questions of mechanics. He corrected and revised Bowditch's translation of Laplace's *Traité de Mécanique Céleste*, and years later Peirce dedicated his own work on analytic mechanics to the "cherished and revered memory of my master in science, Nathaniel Bowditch, the father of

---

[1]See [**129**, pp. 9–14].

American Geometry." Benjamin himself, came to be the most highly regarded American mathematician of his generation.[2] He held the Perkins Chair in Mathematics and Astronomy at Harvard (1842–1880) after having served as University Professor of Mathematics and Natural Philosophy for the previous nine years.

He was a man of broad interests and did not confine himself to mathematics alone. Emerson, Longfellow and Oliver Wendell Holmes were friends of the Peirce family, and their home seems to have been a frequent centre for discussions among the leading scientific figures of Cambridge.[3] As an astronomer, Benjamin took an active part in the foundation of the Harvard Observatory. In fact, the work which first extended his reputation was his remarkably accurate calculations of the perturbations of Uranus and Neptune. Apart from astronomy, another great love of his life was geodesy, for example, establishing a general map of the coastline of the United States entirely independent of detached local surveys. He was director of the longitude determinations of the United States Coast and Geodetic Survey from 1852 to 1867 and Superintendent of the Survey from 1867 to 1874, all the while continuing to serve as professor at Harvard. The United States Coast Survey proved to be not only an additional source of income, but also later provided gainful employment for his son Charles Sanders Peirce. Benjamin himself remained associated with the Survey for the rest of his life, retaining the title of Consulting Geometer.

In 1833 Benjamin married Sarah Hunt Mills, daughter of Elijah Hunt Mills, an eminent lawyer, and had four sons and a daughter. His eldest son James Mills Peirce succeeded his Chair at Harvard and his second son, the aforesaid Charles, was a scientist, semiotician, linguist, philosopher, mathematician and logician. A great committee man, Benjamin Peirce was a member of the American Philosophical Society, an associate of the Royal Astronomical Society, London, and a fellow of the American Academy of Arts and Sciences. In 1847 he was one of a committee of five appointed by this Academy to draw up a program for the organization of the Smithsonian Institution.[4] Peirce was the second American to be elected to the Royal Society of London. His master, Bowditch, had been the first.

---

[2]See [**50**, pp. xiii–xiv].
[3]See [**95**, p. 4].
[4]See [**95**, p. 4].

Although only five feet and seven and three-quarter inches tall, his physical and intellectual presence made a massive impression on students and colleagues alike. Byerly recalls:[5]

> The appearance of Professor Benjamin Peirce, whose long gray hair, straggling grizzled beard and unusually bright eyes sparkling under a soft felt hat, as he walked briskly but rather ungracefully across the college yard, fitted very well with the opinion current among us that we were looking upon a real live genius, who had a touch of the prophet in his make-up.

In an American Mathematical Society Semicentennial address in 1938, George Birkhoff quoted Abbott Lawrence Lowell, former President of Harvard University as follows:[6]

> Looking back over the space of fifty years when I entered Harvard College, Benjamin Peirce still impresses me as the most massive intellect with which I have ever come into close contact, as being the most profoundly inspiring teacher that I have ever had. His personal appearance, his powerful frame and his majestic head seemed in harmony with his brain.

Known as "Benny" to his young students, he encouraged and inspired them. He was full of humor with an abounding love of nonsense and an interest in amateur dramatics. As a teacher, Benjamin Peirce kept abreast of the latest mathematical work in Europe, particularly in England, and used this material as the basis of much of his teaching. He also encouraged his students to undertake original research.[7] Not merely concerned with the operational aspects of the teaching of mathematics, he understood his task as that of advancing the frontiers of mathematics.[8] He was largely responsible for introducing mathematics as a research subject in the United States.[9]

---

[5]See [**25**, p. 5].
[6]See [**11**, p 271].
[7]See [**129**, p. 12].
[8]See [**53**, pp. 8–10].
[9]See [**58**, pp. xix–xx].

His students appreciated his generalizing power, "the quality of his mind which tended to regard any mathematical theorem as a particular case of some more comprehensive one ... so that we were led onward to constantly enlarging truths."[10] He was a profoundly inspiring teacher. At Harvard he produced a series of textbooks "which, while distinctly inferior to the best current in his time, were certainly stimulating."[11] Although inspiring, he was not always easily understood. The following anecdote gives some flavor of his teaching style. At a meeting of, the National Academy of Sciences, Benjamin Peirce spent an hour filling the blackboard with equations only to remark: "There is only one member of the Academy who can understand my work and he is in South America."[12]

Pycior also comments on Peirce as a teacher, "He was not revered for his pedagogical skills: his lectures often degenerated into the furious scribblings of a research mathematician in pursuit of the solution to a fascinating problem; he often refused to answer what he considered ill-posed questions raised by his students. Yet he displayed an enthusiasm for his subject which made a lasting impression on at least the future mathematicians in his classes."[13]

The flavor of a Peirce lecture is captured perfectly by President Emeritus Eliot.[14]

> He was no teacher in the ordinary sense of that word. His method was that of the lecture or monologue, his students never being invited to become active themselves in the lecture room. He would stand on a platform raised two steps above the floor of the room, and chalk in hand cover the slates which filled the whole side of the room with figures, as he slowly passed along the platform; but his scanty talk was hardly addressed to the students who sat below trying to take notes of what he said and wrote on the slates. No question ever went out to the class, the majority of whom apprehended imperfectly what Professor Peirce was saying.

---

[10]See [**4**, pp. 4–5].
[11]See [**6**, pp. 393–398].
[12]See [**50**, pp. xiii–xiv].
[13]See [**167**, p. 541].
[14]See [**51**, p. 2].

In 1862, Thomas Hill then President of Harvard University inaugurated a series of university lectures. These lectures were not to be technical, though advanced. They were to be stimulating as well as informing, and women were encouraged to attend them as well as men. The intellectual requirements of Benjamin Peirce's lectures proved to be too exacting for his audience but his aspect, manner and his whole personality held and delighted them. An intelligent Cambridge matron who had just come home from one of Professor Peirce's lectures was asked by her wondering family what she had got out of the lecture. She answered:[15]

> I could not understand much that he said; but it was splendid. The only thing I now remember in the whole lecture is this: Incline the mind to an angle of 45° and periodicity becomes non-periodicity and the ideal becomes real.

The fact that Benjamin Peirce had considerable influence and persuasive powers amongst his contemporaries is shown in his successful championship of the quaternions. A favorite topic was the *Quaternion Analysis* of W. R. Hamilton. He said, "I wish I was young again, that I might get such power in using it as only a young man can get."[16] He encouraged his students to study this subject. Instruction on the quaternions spread to over ten other American colleges and universities, apparently as a result of Peirce's influence.[17]

Benjamin was a deeply religious man and a committed Unitarian. He often referred to God as "the Divine Geometer" and thought of science as the knowledge of God. He regularly interjected religious observations into his Harvard lectures. W. E. Byerly, a student of his from 1867 through to 1871, and later professor of mathematics at Cornell University, recalled:[18]

> I can see him now at the blackboard, chalk in one hand and rubber in the other, writing rapidly and erasing recklessly, pausing every few minutes to face the class and comment earnestly, perhaps on the results of an elaborate calculation, perhaps on the greatness of the Creator.

---

[15] See [**51**, p. 3].
[16] See [**130**, p. 174].
[17] See [**30**].
[18] See [**25**, p. 5].

More particularly, Byerly noted during one lecture on the quaternions that Peirce claimed that Hamilton's new mathematical system was applicable to the physical world as well as pleasing to the human mind. "The mind of man and that of Nature's God must work in the same channels."[19] Peirce's religious beliefs were, as we shall see, to have an impact on his philosophy of mathematics and on his son Charles Peirce's logic.

To sum up this portrait of the man, his son, Charles, described him as follows:[20]

> ...the leading mathematician of the country in his day, a mathematician of the school of Bowditch, Lagrange, Laplace, Gauss and Jacobi, a man of enormous energy, mental and physical, both for the instant gathering of all his powers and for long-sustained work; while at the same time he was endowed with exceptional delicacy of sensation, both sensuous and sentimental. But his pulse beat only sixty times in a [minute] and I never perceived any symptom of its being accelerated in the feats of strength, agility and skill of which he was fond, although I have repeatedly seen him save his life by a hairbreadth; and his judgement was always sane and eminently cool.

However, there is evidence that Charles was groomed by his father academically at the expense of his personal and social development. Charles always felt that he was in his father's shadow.[21]

> ...he underrated the importance of the powers of dealing with individual men to those of dealing with ideas and with objects entirely governed by exactly comprehensible ideas, with the result that I am today so destitute of tact and discretion that I cannot trust myself to transact the simplest matter of business that is not tied down to rigid forms.

---

[19] See [**25**, p. 6].
[20] See [**95**, p. 4].
[21] See [**50**, p. 4].

2.2. The Publication and Distribution of *Linear Associative Algebra*

Benjamin Peirce also presented a number of papers to the National Academy of Sciences of which he was a founding member. These developed into his book *Linear Associative Algebra* which first appeared in 1870. One hundred lithographed copies were prepared by Julius Erasmus Hilgard (1825–1891) who was Assistant Superintendent at the Coast Survey at the time.[22] Surprisingly it was Charles who having annotated *Linear Associative Algebra* for publication in *The American Journal of Mathematics*, vol. 4 (1881), pp. 97–229, claimed the credit for initiating and promoting the work.[23]

> I had first put my father up to that investigation by persistent hammering upon the desirability of it ... His mind moved with great rapidity and it was with much difficulty that he brought himself to write out even the briefest record of its excursions.

According to Victor Lenzen a manuscript of 1909 has Peirce describing the circumstances at length.[24]

> About 1869 my studies of the composition of concepts had got so far that I very clearly saw that all dyadic relations could be combined in ways capable of being represented by addition (and of course subtraction, by a sort of multiplication ... and by two kinds of involution ... But I found my mathematical powers were not sufficient to carry me further ... I therefore set to work talking incessantly to my father (who was greatly interested in quaternions) to try to stimulate him to the investigation of all systems of algebra which instead of the multiplication table of quaternions ... had some other more or less similar multiplication table. I had hard work at first. It evidently bored him. But I hammered away, and suddenly he became interested and soon worked out his great book on linear associative algebra.

---

[22]See [**72**] for information about the preparation and publication of the 1870 lithographic version.
[23]See [**5**, p. 526].
[24]See [**113**, p. 239].

The original title page of the memoir is reproduced below. In the dedication "To my friends" on the first page of the memoir, the very first sentence reads: "This work has been the pleasantest mathematical effort of my life." It seems to have been a copy belonging to George Davidson and claims that only 50 copies were printed. However a copy of a letter from Hilgard to William Adams Richardson made on 15 March 1871 by Thomas Hill, President of Harvard University from 1862 to 1868 confirms that 100 copies were made.[25]

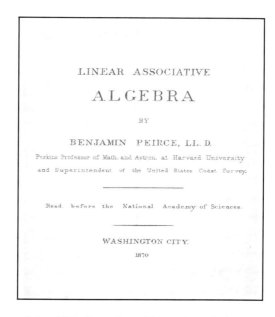

FIGURE 2.1.   Title Page from Linear Associative Algebra

The introduction to *Linear Associative Algebra* concludes with:

> The most laborious part was that of preparing the copy, which was written in lithographic ink on ordinary well sized writing paper. A transfer of these written pages, twelve at a time, was made on a lithographic stone, and 100 copies were printed, after which the transfer was rubbed off, and the next twelve pages laid down on the stone.

[25]See [**72**, p. 605].

Language of algebra	Symbols together with laws for their combination
Art of algebra	Methods of using the symbols in the drawing of inferences
Application of algebra	Interpretation of the symbols alone and in combination

TABLE 2.1. Three Dimensions of Algebra

> The copy was written by a lady who understood not one word of the investigation, but who by great attention succeeded in making a copy far more free from errors that any printers proof ever is, considering Prof. Peirce's chirography it was a wonderful performance.

The *Linear Associative Algebra* is divided into numbered sections, the first of which deals with the definition of mathematics starting with the famous definition "Mathematics is the science which draws necessary conclusions." The different varieties of mathematics are then considered in Section 2, ending with a definition of algebra as *formal mathematics*. Section 3 discusses the distinction between qualitative and quantitative relations while Section 4 is concerned with a three-fold division of algebra shown in Table 2.1.

Benjamin Peirce probably intended to make this the basis of three separate books but only Book 1 on *The Language of Algebra* was ever written. Sections 8 through 13 establish the alphabet of the language. For example, in Section 9 we have:

> The present investigation not usually extending beyond the sextuple algebra, limits the demand of the algebra for the most part to six letters, and the six letters, i, j, k, l, m and n will be restricted to this use except in special cases.

Sections 14 through 24 set up the notation of the algebra under the heading "The Vocabulary" including the symbols > and < for a part-whole theory of classes and symbols for the operations +, − and ×. Benjamin Peirce explicitly states on page 15 of the lithograph that Hamilton's notation of facient, faciend and factum will be used instead of the more common multiplier, multiplicand and product, probably as a compliment to Hamilton. Section 25 introduces for the first time the key ideas of nilpotency and idempotency defined respectively by the equations $A^n = 0$ and $A^n = A$.

Division is defined as the reverse of multiplication but no symbol for division is introduced. Sections 28 through 37 are concerned with *The Grammar* of the algebra, in particular the quantitative forms. In a note on page 19 of the lithograph, Peirce criticizes Hamilton for excluding imaginary numbers from the interpretation of quaternions on the grounds that "like the restrictions of the ancient Geometry, they are inconsistent with the generalizations and broad philosophy of modern science."

In Section 29, the term coefficient is defined as a quantity a such that $Aa = aA$. Sections 30 through 33 define the distributive and associative laws of multiplication and the commutative principle. Section 34 gives the definition of a linear algebra as "an algebra in which every expression is reducible to the form of an algebraic sum of terms, each of which consists of a single letter with a quantitative coefficient."

Benjamin also cites De Morgan's *Triple Algebra*[26] which was clearly an inspiration to him and notes that it adopts the distributive, associative and commutative principles whereas this last principle was emphatically rejected in the *Linear Associative Algebra*. Section 36 defines a symmetrical and cyclic algebra while Section 38 heralds the start of the descriptions of the linear associative algebras. The first important axiom is given in Section 40:

> In every linear associative algebra, there is at least one idempotent or one nilpotent expression.

This axiom is discussed in more detail later in this chapter.

The following sections set up the basis and units of an algebra and their laws of combination, resulting in a multiplication table in Section 46 that classifies the letters of an algebra into one of four distinct groups. Benjamin's general notion of multiplication as shown in these tables is very close to our concept of relative product and is also the definition of multiplication later used by Charles Peirce in his discussion

---

[26]See [**35**].

of elementary relatives in his 1870 paper "Description of a Notation for the Logic of Relatives."[27]

In Section 50, Benjamin Peirce states as the necessary condition for a pure algebra that the four different groups[28] of an algebra should be united by a multiplication relation that links the units of one group to each of the other groups. The properties of units in these groups are then investigated. Several axioms involving idempotency, nilpotency and the order of an algebra and group are then established, some with short proofs. By Section 71, Benjamin Peirce states that "sufficient preparation is now made for the investigation of special algebras." His investigation consists of a methodical calculation of the products of the units of an algebra and the values of their coefficients resulting in a multiplication table for each algebra.

A discussion of the single, double, triple, quadruple, quintuple and sextuple algebras starts from Section 72. The multiplication of the units of each algebra is investigated with single units producing two multiplication tables and therefore two algebras, two units producing three possible algebras, triple units producing five possible algebras with two sub-cases, quadruple units producing eighteen algebras with three sub-cases, the quintuple units producing sixty-five algebras. This makes a total of 163 algebras and six sub-cases.

In conclusion, the three-fold aim of Benjamin Peirce was to

- ○ list all types of number systems in the form of their multiplication tables for systems of up to six units,
- ○ develop a calculus and symbolic method for these systems, and
- ○ draw inferences and deduce applications for these systems.

The *Linear Associative Algebra* of Benjamin Peirce is almost totally confined to the first.

Linear Associative Algebra was first spread by personal contacts, as the friends and colleagues of Benjamin Peirce at the Coast Survey and in the National Academy of Sciences were presented with a lithographic version. Exactly how many of the

---

[27]See [19].

[28]This use is not what we would understand as a group as defined in current group theory but a classification dependent on the unit's multiplication properties.

lithographic copies were produced? Why was this important memoir published in this format and in such a limited edition? Max Fisch provided a possible solution, to wit a purely financial consideration.[29]

> Late in the spring, since the National Academy, only seven years old, had as yet no funds for printing the papers or books its members presented, Julius E. Hilgard, a fellow member of the Academy, took Superintendent Peirce's manuscript, had it copied in a more ornate and legible hand, and then had fifty copies lithographed from it.

This however disagrees with a letter from Hilgard to Robinson, which claims that one hundred copies were printed.[30]

In November 1870, Benjamin Peirce, on his way to Sicily on an expedition to study an eclipse of the sun for the U. S. Coast Survey, gave one copy of the lithograph to the American Ambassador in Berlin who was a personal friend. One copy was also presented to the Berlin Academy. Charles his son presented a copy to Augustus De Morgan, when he visited England with the same expedition under his father's leadership. This copy also contains a charming letter of introduction from Benjamin recommending his son and referring to De Morgan's own work on linear algebra as printed in the Memoirs of the Cambridge Philosophical Society (probably the *Triple Algebras*), stating that *Linear Associative Algebra* was not written without "a careful perusal of the ... treatise."[31]

2.3. Background to *Linear Associative Algebra*

The *Linear Associative Algebra* arises very much from the tradition of the English algebraic school of the early nineteenth century. Continental calculus and mechanics, in particular the works of Laplace, Argobast and Lagrange, had stimulated English mathematicians such as Herschel, Babbage and Peacock to introduce Continental notation and Lagrangian algebraic calculus.[32]

---

[29] See [**56**, p. xxxiii].

[30] See [**72**, p. 605].

[31] See [**73**, p. 38].

[32] Peacock founded the Analytical Society in 1812 at Cambridge University together with Herschel and Babbage. The initial purpose of the Society was to publicise the Leibniz notation of the calculus rather

From 1813 to 1817, Herschel's work on finite-difference equations and series, and Babbage's work on functional equations were based on algebraic abstraction of procedures, and symmetry and used the methods of Lagrange and Laplace.[33] By 1830 George Peacock distinguished between *universal arithmetic* or *arithmetical algebra* and *symbolical algebra* by means of the *principle of the permanence of equivalent forms*. He observed:[34]

> Whatever form is Algebraically equivalent to another, when expressed in general symbols, must be true, whatever these symbols denote.

However De Morgan was to establish his own approach which partly agreed with and partly diverged from, that of Peacock[35], and which relied upon truth and the interpretation of the symbols and of the theories of which they were components. Both De Morgan and Boole continued this tradition of working with symbolical methods (although on different lines). From 1830, Peacock, De Morgan and Gregory looked on algebraic research as the construction of abstract axiomatized systems called symbolic because they were capable of various interpretations. They were aware that it could be impossible to find an interpretation of a given system in mathematics or outside it.[36] Peirce was to follow De Morgan's lead.

Benjamin Peirce did not share the difficulties of the English School because he was sure that such an interpretation was always available in the physical world, believing as he did that such algebras proceeded from God, "the divine Geometer," and so had an expression in Nature. His theological argument justified his researches that led to novel ideas falling outside the normal rules of algebra, without too much regard for use or meaning. As Pycior writes:[37]

---

than the Newtonian. The society helped the communication of the continental mathematics of Laplace, Argobast, and Lagrange to England.

[33]See the first two chapters of [**133**] for the background in French mathematics 1770–1830 and Herschel and Babbage on the calculi of operations and functions 1812–1822.

[34]See [**135**, p. 104].

[35]See [**133**, p. 35].

[36]See [**132**, p. 213].

[37]See [**168**, p. 144].

This argument – what man thought, God thought, and so it was reality – was reiterated again and again in Peirce's writings.

The algebraic logics of De Morgan and Boole also influenced Benjamin to consider only the laws of combination of symbols and not their interpretation. Boole went on to interpret logic as a system of processes which take place with the help of symbols and whose laws are the same as the laws of a system of algebra with the exception of $x = x^2$, the index law of logic which is not generally true for symbols of quantity. Searching for a parallel between the laws of logic and mental processes, Boole first found it necessary to assemble various 'elements of truth' and find fundamental laws, general terms and symbols of these terms to form a language.[38] As shown in Section 2.2, all of this forms an integral part of *Linear Associative Algebra*.

Boole's full title for his book of 1847 is *The Mathematical Analysis of Logic, Being an Essay towards a Calculus of Deductive Reasoning*, and he firmly believed that the sciences are deductive.[39] Benjamin's first sentence in *Linear Associative Algebra*, his famous definition

> Mathematics is the science which draws necessary conclusions

echoes this belief, showing that he also held that sciences have a deductive aspect which involves mathematical processes. The definition is an attempt to broaden the scope of mathematics away from the purely quantitative. Pycior draws attention to Alfred North Whitehead's definition in his *Treatise on Universal Algebra* of 1898 which is strikingly similar to Benjamin's: "Mathematics in its widest signification is the development of all types of formal, necessary, deductive reasoning."[40] Interestingly, Grattan-Guinness discovered two previous drafts of this famous definition:[41]

- Mathematics is the science that draws inferences.
- Mathematics is the science that draws consequences.

and in the cited paper highlights the following two points of interest that arise from the crafting of the final form of the definition.

---

[38]See [**14**, p. 53].
[39]See [**198**, pp. 88–96].
[40]See [**167**, p. 537].
[41]See [**72**, p. 602].

The first of these is the strong link with probability that is also made by Boole and De Morgan. He writes:

> ... to us the word 'necessary' links with possibility; but at that time the closer association would be with probability, which itself treated (among other things) belief and so melded with psychology.

The second is that *Linear Associative Algebra* was written with a notion of the clear distinction between the form and the meaning of an algebraic theory. One consequence of this philosophy was that necessary deductions and conclusions were made from form alone.[42]

Benjamin explained in his introduction to *Linear Associative Algebra* that no law of science could hold without mathematics which deduces from a law all its consequences, and develops them into the suitable form for comparison with observation. From this it can be seen that his definition of mathematics as stated in *Linear Associative Algebra* is then linked with his conception of the match between human thought and mathematical reasoning on the one hand and physical reality on the other, both being manifestations of Divine Laws. Mathematics to him was not humanly devised but was in fact the divine revealer of Truth. As we have seen, so strong was his religious belief that he would interrupt lectures to exclaim upon the existence of God. This link between the laws of mathematics and physical reality is again shown when he was lecturing on his favorite subject, Hamilton's new calculus of quaternions, which he believed was going to be developed into a most powerful instrument of research. He must have been thinking of his *Linear Associative Algebra* for he said that of possible quadruple algebras the one that had seemed to him "by far the most beautiful and remarkable was practically identical with quaternions, and that he thought it most interesting that a calculus which so strongly appealed to the human mind by its intrinsic beauty and symmetry should prove to be especially adapted to the study of natural phenomena."[43]

He did not therefore have to worry about the applicability of his new algebras because his symbolic algebras were to him reflections of the divine Mind and so must have some physical reality. It was clear to him that both Nature and Mathematics originated from God. In short, his acceptance and extension of the symbolical approach

---

[42]See [**73**, pp. 34–35].
[43]See [**4**, p. 6].

to algebra, according to which interpretation was but a secondary consideration, was facilitated by his strong theological belief.

Let us now turn our attention to the most important early influences on the conception of *Linear Associative Algebra*. Sir William Rowan Hamilton's work on complex numbers and his discovery in 1843 of quaternions and the values of the sixty-four constants of multiplication in the system, greatly influenced Benjamin Peirce. However he rejected any philosophical notions that Hamilton attached to the quaternions seeing the system as purely formal. Luboš Nový claims:[44]

> Peirce's interest, which led to his Linear Algebra was aroused from the context from which Hamilton's discovery of the quaternions was generated.

Another important influence was Augustus De Morgan who in 1849 discussed the general commutative and associative systems generated by three units.[45] He also placed complex numbers on a purely symbolic basis. As already mentioned, Benjamin often drew attention to the papers of Hamilton and De Morgan in *Linear Associative Algebra* and this work continues the tradition of the English School: constructing algebraic systems in a limited number of units and listing the various cases; a classic example of which is De Morgan's *Triple Algebras*. This work we have seen mentioned in *Linear Associative Algebra* on page 22 of the lithographic version. However some commentators have thought that this influence did not extend far enough to encompass the often brief and unsatisfactory definitions. Howard Hawkes who published an estimate of *Linear Associative Algebra* in 1902 and a paper on hypercomplex number systems in the Transactions of the American Mathematical Society in 1904 stated:[46]

> It is remarkable that Peirce did not avail himself of the clear and compact definitions of equality and the fundamental operations given by De Morgan and Hamilton.

It can therefore be seen that Benjamin Peirce followed the English School in that his axioms and rules of combinations of his algebras are stimulated by the analogy with ordinary arithmetical algebra. He was concerned with qualitative algebras such

---

[44]See [**132**, p. 218].
[45]See [**35**, pp. 241–254].
[46]See [**88**, p. 95].

as Boole's logical algebra where symbols could be separated from their interpretation, building up a language of algebra. It is clear from *Linear Associative Algebra* that Peirce adopted the symbolical approach of Peacock and the English algebraists. He also emphasized the laws and forms of algebra rather than the meaning of the symbols. The low priority assigned to interpretations and meaning is illustrated by the following anecdote:[47]

> Once after proving a relation in the theory of functions, he dropped his chalk and rubber, put his hands in his pockets, and after contemplating the formula a few minutes turned to his class and said very slowly and impressively, "Gentlemen, that is surely true, it is absolutely paradoxical, we can't understand it, and we don't know what it means, but we have proved it, and therefore we know it must be the truth."

Also, inspired by Hamilton's quaternions, Benjamin Peirce felt free to reject the commutative law in his system of algebras. The algebras developed in *Linear Associative Algebra* are associative and distributive but not necessarily commutative. But on one point Benjamin differed from his hero and went so far as to criticize Hamilton's exclusion of imaginary numbers from his work. By including the possibility of complex coefficients in *Linear Associative Algebra*, Benjamin went still further and sacrificed a determinate division operation. As Fenster states, "Since Peirce insisted that his scalars come from the complex numbers rather than the reals, zero divisors and the indeterminateness of division were potential characteristics of his algebras."[48] This contrasts with today's mathematics where a division operation is defined on all non-zero elements.

Son Charles objected strongly to this weakening of the algebraic analogy.[49]

> There was one feature of this work, however, which I never could approve of, and in vain endeavored to get him to change. It was his making his coefficients, or scalars, to be susceptible of taking imaginary values.

---

[47]See [**25**, p. 6].
[48]See [**54**, p. 76f].
[49]See [**5**, p. 526].

Charles proved to be justified in this position because later developments of linear associative algebras as hypercomplex number systems favored unique division among its elements and allowed its coefficients to come from the real or complex numbers.[50]

Benjamin defended his position in a footnote on page 19 of the lithograph:

> Hamilton's total exclusion of the imaginary of ordinary algebra from the calculus as well as from the interpretation of quaternions will not probably be accepted in the future development of this algebra. It evinces the resources of his genius that he was able to accomplish his investigations under these trammels.

Charles however, was not impressed and called this footnote "pure bosh."[51] This shows how far Benjamin was prepared to go to achieve his vision of a general and broad based approach to algebra that moved ever further from quantitative or arithmetical algebra.

If we are to believe Charles Peirce, it may be that the overriding influence that inspired *Linear Associative Algebra* was Charles's desire to seek some application of his algebraic logic. Since his father's linear algebras could be represented as relative terms, this gave a clear justification to his own logic. So it may be that although Benjamin did not pay much regard to the interpretation and use of his algebras, his son Charles did. Although the majority of the lithographic copies of *Linear Associative Algebra* went to American friends and this necessarily prevented wider access to the work, it was then published after Benjamin's death in a new edition with addenda and notes by Charles in volume four of the *American Journal of Mathematics* of 1881 and reprinted in 1882 in book form by Van Nostrand. It is now recognized as Benjamin Peirce's finest work, and considered to be the first major original contribution to mathematical progress in the United States. Raymond Clare Archibald claims: "There seems to be no question that his *Linear Associative Algebra* was the most original and able mathematical contribution which Peirce made ... " In his Synopsis of *Linear Associative Algebra* published by the Carnegie Institution in 1907, J. B. Shaw characterized the work as "really epochmaking."[52]

---

[50]See [**54**, p. 77].

[51]See [**5**, p. 526].

[52]See [**189**, p. 6].

Secondly, and more controversially, he claimed that Benjamin Peirce wished, like Boole and De Morgan, to lay the foundations for mathematics with some kind of symbolic logic.[53] This claim is puzzling because it was in fact his son Charles Peirce who developed an algebraic logic although not for the purpose of providing the foundations of mathematics, and Benjamin always counseled his son away from logic as he claimed it was neither profitable nor useful. Although Nový placed *Linear Associative Algebra* very much as a successor to the work of the English algebraists, it does differ in one respect. Helena Pycior has commented on the fact that *Linear Associative Algebra* breaks away from one of the fundamental principles of the earlier English algebraists – that of Gregory's principle of the permanence of equivalent forms in which it is stated that it is the laws of arithmetic which dictate the laws of symbolic algebra.[54] This freedom from the conventional arithmetic laws led the way to the development of a number of different algebraic systems. First Hamilton with his non-commutative quaternions violated this principle, and then in 1844 in a paper read before the Cambridge Philosophical Society on triple algebras, De Morgan developed a few non-associative triple algebras.[55] Pycior correctly states that *Linear Associative Algebra* was a pioneer work in this tradition, both in American mathematics and in current abstract algebra. Pycior also states:[56]

> Because of *Linear Associative Algebra* ... Benjamin Peirce deserves recognition, not only as a founding father of American mathematics, but also as a founding father of modern abstract algebra.

2.4. Analysis of the Algebras

2.4.1. Definitions and Axioms of the Algebras. Let us turn our attention to the definitions and axioms of *Linear Associative Algebra*. In this section after setting out a number of axioms introducing the terms and units of the algebras and a string of definitions including those of idempotency and nilpotency, we then consider the most important operation in *Linear Associative Algebra*, that of multiplication. The relevant axioms and definitions are outlined below. The numbered brackets correspond to the relevant formula in *Linear Associative Algebra*. We shall concentrate in particular

---

[53]See [**132**, p. 226].
[54]See [**167**].
[55]See [**35**].
[56]See [**167**, p. 551].

on a selection of those axioms that are necessary for the deduction of many cases of particular algebras. Note that in the following definitions Benjamin Peirce is treating multiplication as an operation, i.e., as T operating on A or A operating on T.

All algebras treated in *Linear Associative Algebra* are linear associative algebras where a linear algebra is defined as "an algebra in which every expression is reducible to the form of an algebraic sum of terms, each of which consists of a single letter with a quantitative coefficient."[57] A linear associative algebra is a linear algebra in which the associative principle of multiplication is adopted and "extends to all the letters of its alphabet."[58] It is interesting to note that nowhere in *Linear Associative Algebra* did Benjamin Peirce define equivalent algebras. However he does discard some algebras that are *virtual repetitions* of others. These are the cases where the second algebra produces a multiplication matrix that is a transpose matrix of the first algebra with rows being transposed to columns.

Definition 22: The sign × may be adopted from ordinary algebra with the name of the sign of *multiplication* but without reference to the meaning of the process. The result of multiplication is to be called the *product*. The terms which are combined by the sign of multiplication may be called *factors*; the factor which precedes the sign being distinguished as the *multiplier* and that which follows it being the *multiplicand*. The words multiplier, multiplicand and product, may also be conveniently replaced by the terms adopted by Hamilton, of *facient*, *faciend*, and *factum*. Thus the equation of the product is

$$\text{multiplier} \times \text{multiplicand} = \text{product}$$
$$\text{or}$$
$$\text{facient} \times \text{faciend} = \text{factum}.$$

When letters are used, the sign of the multiplication can be *omitted* as in ordinary algebra.

Definition 23: When an expression used as a factor in certain combinations gives a product which vanishes it may be called in those combinations a *nilfactor*. Where it produces vanishing products as the multiplier, it is *nilfacient*, but where it is the multiplicand of such a product it is *nilfaciend*.

---

[57]See [**137**, p. 22].
[58]See [**137**, p. 25].

Definition 24: When an expression used as a factor in certain combinations, over-powers the other factors and is itself the product, it may be called an *idemfactor*. When in the production of such a result, it is the multiplier, it is *idemfacient*, but when it is the multiplicand, it is *idemfaciend*.

Definition 25: When an expression which is raised to the square or any higher power, vanishes, it may be called nilpotent; but when raised to a square or higher power, it gives itself as the result it may be called idempotent.

The defining equation of nilpotent and idempotent expressions are respectively $A^n = 0$, and $A^n = A$; but with reference to idempotent expressions, it will always be assumed that they are of the form

$$A^2 = A,$$

unless it be otherwise distinctly stated.

It should be noted in the following examples that only those algebras in which every expression can be expressed as a linear combination and which obey the associative law of multiplication, are considered. Two important proofs in *Linear Associative Algebra* are outlined overleaf.

Definition 40: In every linear associative algebra, there is at least one idempotent or one nilpotent expression.

Let A denote any combination of letters. Its square is generally independent of A and its cube may be. Then the number of powers of A which are independent of A and of each other cannot exceed the number of letters of the alphabet, so there must be some least power of A which is dependent upon the inferior powers.

$$\sum_m a_m A^m = 0$$

By this Benjamin Peirce means

$$a_k A^k = a_1 A^1 = a_2 A^2 + a_3 A^3 \ldots a_{k-1} A^{k-1},$$

and this can be expressed as an equation in the form of a linear combination of powers of A equal to 0

(2.1) $$a_1 A^1 = a_2 A^2 + a_3 A^3 \ldots a_m A^m = 0.$$

Benjamin Peirce continues

$$a_1 A + BA = 0,$$

where BA is an algebraic sum of the square and higher powers of A. This means that B is itself an algebraic sum of powers of A, i.e.,

$$B = a_2 A^1 + a_3 A^2 + a_4 A^3 \ldots a_m A^{m-1},$$

so

$$(B + a_1) = 0$$

and so successively, multiplying by powers of A,

$$(B + a_1)A^m = 0.$$

Hence

$$(B + a_1)B = 0.$$

Benjamin Peirce is justified in this step because from (2.1) A^m can be expressed as a linear combination of "inferior" powers from A^1 to A^{m-1}, i.e., as B. Two brief equations follow:

(2.2) $$B^2 + a_1 B = 0$$

(2.3) $$\left(\frac{-B}{a_1}\right)^2 = \frac{-B}{a_1}$$

Let us try to follow the argument by supplying the missing stages. Assume that a_1 is non-zero. Dividing successively by a_1 we get:

$$\frac{B^2}{a_1} + B = 0 \text{ and } \left(\frac{B}{a_1}\right)^2 + \frac{B}{a_1} = 0,$$

$$\left(\frac{B}{a_1}\right)^2 = -\frac{B}{a_1} \text{ and } \left(-\frac{B}{a_1}\right)^2 = -\frac{B}{a_1}$$

so

$$-\frac{B}{a_1}$$

is an idempotent expression. But if a_1 vanishes this expression becomes infinite and so from 2.2 we have

$$B^2 = 0$$

and therefore B is a nilpotent expression.

Following on from Definition 40, let us assume there is an idempotent expression.

Definition 41: When there is an idempotent expression in a linear associative algebra, it can be assumed to be one of the independent units[59] and be represented by one of the letters and called the basis.[60]

The remaining units can be separated into four distinct groups in a classificatory approach, with respect to this basis, as can be seen in Table 2.2.

1	2	3	4
$iA = A$ $Ai = A$	$iA = Ai$ Idemfaciend	$Ai = A$ Idemfacent	$iA = 0$ $Ai = 0$
Idemfactors	$Ai = 0$ Nilfacient	$iA = 0$ Nilfaciend	Nilfactors
dd	dn	nd	nn

TABLE 2.2. Classification of the remaining units A in terms of the basis i

The remaining units A are put into four groups dd, dn, nd or nn according to their relation to the basis unit i.

Here A is expressed in two parts. The first letter gives its name as a faciend, the second giving its name as a facient. The two letters are d and n, of which d stands for idem and n for nil.

All remaining units are either idemfaciend or nilfaciend. We have $i^2 = i$. The product by the basis of another expression such as A may be represented by B, so that $iA = B$. Thus,

$$iB = i^2A = iA = B$$

so B is idemfaciend. In addition,

$$i(A - B) = iA - iB = B - B = O$$

so $A - B$ is nilfaciend. Therefore A is made up of two parts, one of which is idemfaciend, the other nilfaciend but either of these parts may be wanting and we have A wholly idemfaciend or wholly nilfaciend.[61]

---

[59]Consider these as letters i, j, k, ... making up the algebra.
[60]Here $i^2 = i$. .
[61]See [**137**, p. 28].

In this ingenious explanation it is not clear that Benjamin Peirce intends A to represent the class of possible expressions generated by units of the algebra, where B is a subclass of A, produced by multiplication by i. So we can get two groups of idemfaciends and nilfaciends. It can be similarly shown that all the remaining units are either wholly idemfacient or nilfacient.

Idemfaciend		Nilfaciend	
Idemfacient	Nilfacient	Idemfacient	Nilfacient
1	2	3	4

TABLE 2.3. Idemfaciend/Nilfaciend

Next Benjamin Peirce builds a table showing the products of expressions in these four groups. To do this he defines a factorially homogeneous expression as an algebraic sum of letters of a group that belongs to the same group, and continues:

Definition 43: The product of two factorially homogeneous expressions, which does not vanish, is itself factorially homogeneous, and its faciend name [or nature] is the same with that of its facient [part], while its facient name is the same as that of its faciend [part].

Thus if A and B are, each of them, factorially homogeneous they satisfy the equations

$$i(AB) = (iA)B,$$
$$(AB)i = A(Bi),$$

which shows that the nature of the product as a faciend is the same with that of the facient A, and its nature as a facient is the same with that of the faciend B.

Let us write out Benjamin Peirce's products explicitly. We have, remembering that the two-letter notation gives the nature of the expression first as faciend then as facient:

$$dd \times dd = dd \quad nd \times dd = nd$$
$$dd \times dn = dn \quad nd \times dn = nn$$
$$dn \times nd = dd \quad nn \times nd = nd$$
$$dn \times nn = dn \quad nn \times nn = nn$$

Definition 45: Every product vanishes of which the facient is idemfacient, while the faciend is nilfaciend; or of which the facient is nilfacient while the faciend is idemfaciend.

For in either case, this product involves the equation

$$AB = (Ai)B = A(iB) = 0.$$

Benjamin Peirce intends here a product $A \times B$ of either the form

idemfacient × nilfaciend $*d \times n* = 0$

or

nilfacient × idemfaciend $*n \times d* = 0$

where $*$ stands for either n or d. Considering these zero products we have:

$$dd \times nd = 0 \quad nd \times nd = 0$$
$$dd \times nn = 0 \quad nd \times nn = 0$$
$$dn \times dd = 0 \quad nn \times dd = 0$$
$$dn \times dn = 0 \quad nn \times dn = 0$$

Definition 46: The combination of the propositions of §43 and §45 is expressed in the following form of a multiplication table.

		Faciend		
		dd	**dn**	**nd**	**nn**
	dd	dd	dn	0	0
Facient	**dn**	0	0	dd	dn
	nd	nd	nn	0	0
	nn	0	0	nd	nn

TABLE 2.4. Faciend/Facient I

Equivalently as multiplication between groups 1,2,3 and 4 we have:

		Faciend		
		1	**2**	**3**	**4**
	1	1	2	0	0
Facient	**2**	0	0	1	2
	3	3	4	0	0
	4	0	0	3	4

TABLE 2.5. Faciend/Facient II

Definition 47: It is apparent from the inspection of this table, that every expression which belongs to the second or third groups is nilpotent.

Definition 50: Since the products of the units of a group remain in the group, they cannot serve as the bonds for uniting different groups, which are the necessary conditions of a pure algebra. Neither can the first and the fourth groups be connected by direct multiplication because the products vanish. The first and fourth groups therefore, require for their indissoluble union into a pure algebra that there should be units in each of the other two groups.

Let us now consider some examples of the theorems of *Linear Associative Algebra* that will be needed to establish some of the algebras to be considered later in this section.

Theorem 51: In an algebra which has more than two independent units, all the units except the base cannot belong to the second or the third group. For in this case, each unit taken with the base would constitute a double algebra, and there could be no bond of connection to prevent their separation into distinct algebras.

Theorem 57: In a group or an algebra which has no idempotent expression, all the expressions are nilpotent.

A nilpotent group or algebra may be said to be of the same order with the number of the powers of its basis that do not vanish, if the basis is selected which has the greatest number of powers which do not vanish.

Theorem 59: In a group or an algebra which contains no idempotent expression, any nilpotent expression may be selected as the basis; but one is preferable which has the greatest number of powers which do not vanish.

All the powers of the basis which do not vanish may be adopted as independent units and represented by the letters of the alphabet.

Theorem 60: It is obvious that in a nilpotent group, which is of the same order with the number of letters which it contains, all the letters except the basis are the successive powers of the basis.

Theorem 63: In every nilpotent group, the facient order of any letter which is independent of the basis can be assumed to be as low as the number of letters which are independent of the basis.

It also holds that there is always a value of A_1 which will give $i^m A_1 = 0$. The following theorems are then developed and accompanied by short proofs:

Theorem 64: In a nilpotent group, the order of which is less by unity than the number of letters, the letter which is independent of the basis and its powers may be selected, so that its product into the basis, shall be equal to the highest power of the basis which does not vanish, and its square shall either vanish or shall <u>also</u> be equal to the highest power of the basis which does not vanish.

It follows from (60) that the algebra consists of successive powers of the basis i and the independent unit j. The order is n and so i^{n+1} and higher powers of $i = 0$. For example suppose let us take the order of i to be three so that $i^4 = 0$: we have four letters i, i^2, i^3 and j with i as the basis and j as the independent letter: $i^4 = 0$ then we require $ji = i^3$, and $j^2 = 0$ or $j^2 = i^3$. From (63) since there is only one independent letter j, the facient order of i will be 1 and so we have $ij = 0$ which gives

(2.4) $$i ji = i j^2 = 0.$$

Here is the proof exactly as it appears on page 41 with a,b,a' and b' representing the coefficients of the lithographic version. Benjamin Peirce has:

$$ji = ai^n + bj \tag{1}$$

$$j^2 = a'i^n + b'j \tag{2}$$

$$0 = ji^n + 1 = bji^n = b^n ji = b \tag{3}$$

$$ji = ai^n \tag{4}$$

$$j^2i = aji^n = 0 = b'j^2 2 = b' \tag{5}$$

$$j^2 = a'i^n \tag{6}$$

Let us consider these five equations in more detail. (1) and (2) express the product of the elements of the algebra as a linear combination of the units of the algebra and we are assuming that ji and j^2 are non-zero, otherwise one condition of the theorem is then proved. In the case of ji and j^2 being non-zero, since (2.4) $i(ji) = 0$ and $ij^2 = 0$, we cannot have ji and j^2 expressed as powers of i less than n otherwise (2.4) would not hold, so that Benjamin Peirce is justified in expressing ji and j^2 as $ai^n + bj$ and $a'i^n + bj$ respectively.

Multiplying (1) by i^n from the right and substituting from (1) we have

$$ji^{n+l} = jl(i^n) = (ai^n + bj)i^n = bji^n,$$

since powers of i greater than n vanish as the group is nilpotent.

Consider the equation $ji^{n+1} = bji^n$. This equation holds for all values of n so substituting in $n = 1$ we get:

$$ji^2 = bji$$

and

$$ji^3 = (ji)i^2 = (ai^n + bj)i^2 = bji^2$$
$$= b(bji) = b^2 ji.$$

and similarly

$$ji^4 = (ji)i^3 = (ai^n + bj)i^3 = bji^3$$
$$= b(b^2 ji) = b^3 ji.$$

In this way, Benjamin Peirce is justified to claim $bji^n = b^n ji$ in (3). Since ji is non-zero we have $b = 0$. Substituting $b = 0$ in (1) gives us $ji = ai^n$ or (4). Therefore multiplying by j from the left:

$$j^2 i = a ji^n.$$

However, multiplying (2) by i from the right and noting higher powers of i vanish, we obtain the equation

$$j^2 i = (a'i^n + b'j)i = b'ji.$$

Since ji is non-zero we have $b' = 0$. Notice that I disagree with Peirce here who has $b'j^2$ instead of $b'ji$ in (5). Substituting in $b' = 0$ in (2) we obtain (6). Of the remaining theorems the following two are used in producing the algebras:

Theorem 67: In the first group of an algebra, having an idempotent basis, all the expressions except the basis may be assumed to be nilpotent.

Theorem 69: ...if the idempotent basis were taken away from the first group of which it is the basis, the remaining letters of the first group would constitute by themselves a nilpotent algebra.

It is at this stage that Benjamin Peirce is able to begin an investigation of special algebras starting with single algebras through sextuple algebras using the letters i, j, k, l, m and n and the numbers and coefficients assigned to them according to Table 2.6.

So, for example,

$$jl = a_{24}i + b_{24}j + c_{24}k = d_{24}l + e_{24}m + f_{24}n$$

since j and l have assigned numbers 2 and 4 respectively. For squares, only one number is needed:

$$k^2 = a_3 i + b_3 j + c_3 k + d_3 l + e_3 m + f_3 n.$$

i	j	k	l	m	n
1	2	3	4	5	6
a	b	c	d	e	f

TABLE 2.6. Assigning Coefficients

Benjamin Peirce's investigation consists in finding the values of the coefficients a, b, c, d, e and f corresponding to every variety of linear algebra and arranging the resulting products in a multiplication table. The basis is denoted by i. In each algebra the procedure followed is to take i idempotent and i nilpotent.

Each of these are then split into subcases to develop all the possible algebras, discarding those in which each letter is not linked by multiplication to each of the other letters, as in those cases no pure algebra results.

2.4.2. The Triple Algebras. We will now confine ourselves to the triple and quadruple algebras as the single and double algebras are fairly straightforward and the quintuple and sextuple algebras are developed in the same way. Benjamin Peirce's method i.e. looking at all the possible products of the units of the algebra and producing multiplication tables for the 'pure' algebras can be seen to be used in each case and the reasoning behind his results is expanded upon.

Table 2.7 is a simple example of a double algebra: This algebra with two units is

(c_2)	i	j
i	j	0
j	0	0

TABLE 2.7. Algebra c_2

defined by the equation $i^2 = j$. All the other products give zero.

2.5. Benjamin Peirce's Addendum to *Linear Associative Algebra*

Benjamin Peirce went on to produce a paper entitled 'On the Uses and Transformations of Linear Algebra'. This was presented to the American Academy of Arts and Sciences on May 11th 1875 and published in the Proceedings.[62] It was then reprinted posthumously as the first part of a series of addenda to the *American Journal of Mathematics* article of 1881, pages 216–221, the other two parts comprising "On the Relative Forms of the Algebras" and "On the Algebras in which Division is Unambiguous," both written by his son Charles Peirce.

In this article Benjamin Peirce addresses the difficult problem of the interpretation and therefore application of his linear algebra. However he argues that interpretation is not always necessary and that the success of the algebra comes from 'its want of significance'. In fact he goes on to say 'the interpretation is a trammel to the use.'[63]

The units or symbols of the linear algebras namely i, j, k, etc. are compared to the imaginary number i, the square root of -1, with its signification of the representative of perpendicularity in quaternions. He then draws upon Hankel numbers, with Clifford's applications to determinants, as being a natural generalization from §63. Charles Peirce, here adds a footnote to the effect that Hankel numbers were given much earlier under the name of clefs by Cauchy and even earlier by Grassmann. That class of algebra which follows Benjamin Peirce's laws of multiplication[64], given the name of quadrates by Clifford, is then discussed. Charles in particular was eager to develop this particular linear algebra in so far as it related to his own algebraic logic. The definition of quadrates proposed by him is presented in the form of a list of relations written as ordered pairs.

If the letters A, B, C, ... represent absolute quantities, differing in quality then the units of the algebra may represent the relations of these quantities, and may be written

$$(A:A)(A:B)(A:C)\ldots(B:A)(B:B)\ldots(C:A)$$

---

[62]See [**138**].

[63]See [**137**, p. 216].

[64]See the tables in section 2.6.

subject to the equations

$$(A : B)(B : C) = (A : C)$$
$$(A : B)(C : D) = 0.$$

Benjamin Peirce then defines unity as being the sum of the units $(A : A)$, $(B : B)$, $(C : C)$, etc., before defining a unit of inversion as a unit which differs from unity but of which the square is equal to unity and a unit of semi-inversion as one of which the square is a unit of inversion. He then draws an analogy with Hamilton's quaternions where all units apart from unity, are units of semi-inversion, and Clifford's biquaternions where all units are units of inversion.

Although *Linear Associative Algebra* deals primarily with associative algebras, the uses of commutative algebras are considered here, with one particular application – the integration of differential equations of the first degree with constant coefficients. Benjamin Peirce highlights the algebra with (a_3), as being useful in providing solutions for Laplace's equation for the potential of attracting masses where the units must satisfy the equation,

$$i^2 + j^2 + k^2 = 0,$$

as is the case with (a_3).

So Benjamin Peirce sought interpretations for his linear algebras in a number of ways. These include drawing on his original inspiration that sprang from Hamilton's quaternions, and recent developments in algebra such as Clifford's biquaternions, and also on the work on composition of relations arising from the algebraic logic of his son Charles Peirce.

2.6. The Reception of *Linear Associative Algebra*

Benjamin Peirce was able in January 1871, after the completion of the expedition, to address the London Mathematical Society on the methods he had used in *Linear Associative Algebra* and to present a copy to the Society. A year later, its President W. Spottiswoode read the paper "Remarks on Some Recent Generalizations of Algebra" on the principal ideas of *Linear Associative Algebra* to the London Mathematical Society, which was published in its *Proceedings*.[65] Arthur Cayley in an address to the

---

[65]See [**192**].

British Association for the Advancement of Science in 1883, praises *Linear Associative Algebra* as a valuable memoir and Peirce's linear algebras as the "analytical basis, and the true basis" of complex numbers, quaternions, and other such algebras. Cayley also appreciated the novelty of Peirce's work, in particular its philosophical treatment of mathematics with its freedom from conventional algebraic rules. However in a telling comment he classified it as "outside of ordinary mathematics."[66]

Linear Associative Algebra had a disappointing reception in Germany. The journal *Jahrbuch über die Fortschritte der Mathematik* announced its intention of publishing *Linear Associative Algebra* but never did and Benjamin's own initiative in presenting a copy to the American Embassy in Berlin for distribution amongst German mathematicians remained without response.[67] Pycior suggests a possible reason when quoting Jekuthiel Ginsburg who attributes the poor reception to the "general distrust shown by European scholars of the achievements of American writers."[68] Nový also attributes the lack of influence of *Linear Associative Algebra* on contemporaries to the generality of the work (they found it difficult to classify as mathematics, describing it instead as philosophy) and its distance from other mathematical problems of the time. Pycior however, explains the lack of impact as due to Peirce's superior mathematical ability. One exception to the general disinterest was *Linear Associative Algebra*'s favourable reception in England.[69] It is not surprising in view of the fact that *Linear Associative Algebra* was inspired by English algebraists that it had such a success. However, when appraising *Linear Associative Algebra* in 1902, Howard E. Hawkes acknowledged that the work had come to attract wide and favourable comment in England and America and attributed the continuing lack of interest from French and German mathematicians as arising from the "misunderstanding of Peirce's definitions, the arbitrariness of Peirce's principles of classification and to Peirce's vague and in some cases unsatisfactory proofs."[70]

In his "Estimate of Peirce's Linear Associative Algebra," Hawkes sought to redress the neglect of *Linear Associative Algebra* and provide the credit which the work deserved. He regarded the *Linear Associative Algebra* as the first systematic attempt to classify and enumerate hypercomplex number systems. After initial praise *Linear*

---

[66]See [**26**, pp. 457–458].

[67]See [**132**, p. 227].

[68]See [**167**, p. 545].

[69]See earlier remarks by Spottiswoode and Cayley.

[70]See [**88**, p. 87].

Associative Algebra was ignored largely because, according to Hawkes, of Peirce's arbitrary methods of classification and his vague, unsatisfactory proofs. Hawkes traces the development by Hamilton and De Morgan of number systems of two and four units in a purely symbolic form. The problem encountered by Benjamin Peirce was a) to devise a classificatory system for number systems and b) to enumerate them. He solved this by grouping together systems with the same number of units. In his classificatory approach he narrowed his field to only pure algebras where the system cannot be divided into two or more separate groups which are not linked by any multiplication relation. It is Hawkes' opinion that this narrow approach led to criticism from Study and Scheffers who also worked in this field. Eduard Study (1862–1930) was Professor of Mathematics at Göttingen and Bonn. He worked in real and complex algebras of lower dimensions. He also visited the United States in 1893 where he taught at several universities but he was mainly based at the Johns Hopkins University. Georg Scheffers (1866–1945) held the Chair of Mathematics in Charlottenburg. Lie greatly influenced Scheffers' work including that on complex number systems.

Hawkes now identified the problem that Benjamin Peirce set out to solve and related the work to the number systems of Study and Scheffers. He wrote:

> Peirce's definitions of pure and mixed algebras correspond exactly to the definitions of irreducible and reducible number systems used by Scheffers, except that the groups in a reducible system can contain no common units.[71]

Hawkes goes on to highlight the five general principles used by Benjamin. These general principles show that although he closely followed the work of English algebraists, he also introduced many original concepts that influenced later mathematicians. The five general principles as explained by Hawkes are:

Classification of Algebras by Units: Systems with the same number of units are classified into one group.

Equivalence: Two systems with the same number of units are considered equivalent if each unit of one system can be expressed as a linear combination of units of the other system, i.e., two systems with the same number of units $e_1, e_2, \ldots, e_n$, and $e'_1, e'_2, \ldots,$

---

[71] See [**88**, p. 92].

e'_n, are considered equivalent if linear relations exist of the type

$$e'_k = \sum_{i=1}^{n} a_{ki} e_i$$

for $k = 1, 2, \ldots, n$.[72]

Pure Systems: A key concept of *Linear Associative Algebra* is that of 'pure algebras'. Peirce's definition of a pure algebra as one in which there is a non-zero product linking each distinct unit is explained by Hawkes in the following way: A system is not pure if its units may be divided into two or more groups (which may have common units), such that the product, if non-vanishing, of two units of the same group is in that group, while the product of units which are not found in the same group is zero. For example:

	i	j	k
i	i	j	0
j	0	0	0
k	k	0	0

TABLE 2.8. Example of a Non Pure System

Here the groups are $\{i, j\}$ and $\{i, k\}$. A pure algebra is one which cannot be divided into such groups. These definitions of pure and nonpure algebras correspond closely to the definitions of irreducible and reducible number systems.

Reciprocity: Peirce assumed throughout his memoir that reciprocal systems in the sense of matrix transposition, are virtual repetitions of each other and are therefore equivalent.

Idempotent Numbers: Idempotent units as defined in *Linear Associative Algebra* lead to the concept of a system of module. If the modulus of a system such as the number μ is defined such that $x\mu = \mu x = x$ for every number x in the system, then the existence of an idempotent number is the necessary, although ot sufficient condition to the existence of a modulus.

---

[72]This is Hawkes's notation and explanation which are not to be found in the original of *Linear Associative Algebra*. See [**88**, p. 91].

This classification enabled Hawkes to clarify the aim of *Linear Associative Algebra*.[73]

> We can now state precisely the problem that Peirce set for himself. He aimed to develop so much of the theory of hyper-complex numbers as would enable him to enumerate all inequivalent, pure, non reciprocal number systems in less than seven units.

In contrast Henry Taber's paper on hypercomplex number systems[74] attempted to place Peirce's methods, i.e., his proofs, on a rigorous basis using only algebraic methods. Taber obtained his results by extending hypercomplex numbers to a more general expression with any domain for the co-ordinates in terms of scalar function theory. In fact Taber criticizes Hawkes in his attempt for using the theory of transformation groups and so introducing unnecessary and foreign methods for the establishment of Peirce's own purely algebraic system. Although Taber called his methods "algebraic"[75] refers to them as "algebra-theoretic" since, in modern mathematics, the term "algebraic" certainly includes group theory.

Taber also claimed that some of the proofs of *Linear Associative Algebra* were invalid. He then supplied such proofs by using scalar function theory which expresses hypercomplex number systems $(e_1, e_2, \ldots, e_n)$ in the form of linear combinations of independent units with scalar coefficients; scalar being taken to represent 'a real or ordinary complex number'[76] In this he was following in the footsteps of Hawkes who used the theory of transformation groups to prove the theorems of *Linear Associative Algebra*. But as we have already mentioned, Taber criticised Hawkes for using group theory techniques and "introducing conceptions unnecessary for the establishment of Peirce's theory and foreign to his methods."[77]

Synopsis of Linear Associative Algebra was published by James Byrnie Shaw, Professor of Mathematics in the James Millikin University in 1907. This work traced

---

[73]See [**88**, pp. 91–94].

[74]See [**195**].

[75]See [**54**, p. 94].

[76]Charles Peirce argued against such use of the term 'scalar', since he reasoned complex numbers have two dimensions and were therefore properly represented by double algebras. The other argument he used was that the use of complex coefficients results in zero divisors and a non-determinate division operation.

[77]See [**195**, p. 510].

the developments of such algebras post *Linear Associative Algebra*. Two such lines are singled out: group theory and matrix theory, and many of the developments are given in terms of one of the above theories. He wrote:[78]

> The first is by use of the continuous group. It was Poincaré who first announced this isomorphism. The method was followed by Scheffers, who classified algebras as quaternionic and nonquaternionic ... The other line of development is by using the matrix theory. C. S. Peirce first noticed this isomorphism, although in embryo it appeared sooner. The line was followed by Shaw and Frobenius. The former shows that the equation of an algebra determines its quadrate units, and certain of the direct units; that the other units form a nilpotent system which with the quadrates may be reduced to certain canonical forms.

Although the Synopsis is a thorough review of the results arising from the theory of linear associative algebras, it does not treat Benjamin Peirce's own methods, concentrating instead on the latest developments by other mathematicians. Shaw reformulated the general principles of algebraic systems in terms of hypercomplex numbers so that a hypercomplex number α can be defined as $\alpha = \sum a_i e_i$ where there is a finite set of r units $e_1, \ldots, e_r$ and coefficients $a_1, \ldots, a_r$. He does this by using the techniques of matrix theory such as determinants, invariants and orthogonality to produce theorems and definitions of a high level of complexity. Shaw attempted to draw together the work of contemporary mathematicians and to this end redefined hypercomplex numbers with terms drawn from Cayley, Frobenius, Scheffers, and Study. For example on page 18 we have the following definition in terms of Bertrand Russell's logical constants:

> A hypercomplex number is an aggregate of r one-many relations, the series of real numbers being correlated with the first r integers. Thus, to the r integers we correlate $a_1, a_2, \ldots, a_r$, all in the range of real numbers. This correlation is expressed by the form
>
> $$a_1 e_1 + a_2 e_2 + \ldots + a_r e_r.$$

---

[78]See [**189**, p. 6].

Nový also ignored Benjamin Peirce's own algebraic techniques and methods but outlined the main sections of *Linear Associative Algebra*, dealing rather with the philosophy and definitions that make up the first section rather than analyzing the methods and results. Nový related *Linear Associative Algebra* with the English algebraic school of 1830 of Peacock, and Gregory, through to the algebraic systems invented by Boole and De Morgan in the 1850's culminating in the work of Sylvester and Cayley. He cited two points of comparison. Firstly *Linear Associative Algebra* is an elaboration of partial algebraic systems with an extensive listing of various cases of such algebras - methods commonly used by representatives of the English algebraic school.

Pycior states the case for Benjamin Peirce to be seen as a pioneer who, following the English algebraist Peacock and the Irish algebraist Hamilton, sought to devise number systems that broke free from the trammels of arithmetic.[79] In particular, Peirce's use of complex coefficients for his own algebras led to zero divisors, i.e., nonzero elements a and b such that $a \times b = 0$, which results in an indeterminate division operation, so not only was the commutative law relinquished but also arithmetic division as well. Pycior attributes the ultimately poor reception of *Linear Associative Algebra* to the advanced nature of the work and the lack of mathematicians of note in America at that time, over and above any failings of presentation and clarity on the part of Benjamin himself as mentioned previously in [**88**]. The novelty and breadth of *Linear Associative Algebra* led to it being classified philosophy or what we would now regard as philosophy of mathematics. Even Cayley who appreciated *Linear Associative Algebra* felt it necessary to stress the novelty of Peirce's results by classifying the algebras in *Linear Associative Algebra* as "outside of ordinary mathematics."[80]

---

[79]See [**167**].
[80]See [**26**, p. 457].

CHAPTER 3

English Influences on Charles Peirce's Algebraic Logic

3.1. Introduction

In a definition similar to that given by Benjamin Peirce for mathematics, Leibniz defined logic as the science "that teaches other sciences a method for discovering and proving all corollaries following from given premises."

Noticing that logic with its terms, propositions and syllogisms bore a resemblance to algebra with its letters, equations and transformations, he tried to present logic as a calculus or a universal mathematics. The two main themes of his thought were:

a) a universal language – an alphabet of thought – in which each symbol represented a simple concept with a combination of symbols for a compound idea, and

b) a calculus of reasoning based on this language to express the main relations between scientific concepts.

Leibniz saw this as a mathematical procedure but more general than existing mathematical methods. The disadvantage of his calculus was that he chose the complicated scheme of assigning numbers to concepts so that the number became the symbol of the concept. The concepts, each assigned a different number, are divided into classes so that Class 1 contains all elementary concepts, i.e., those not further analyzable, and Class 2 contains those concepts definable in terms of those in Class 1, etc.[1]

The long period of time before Leibniz's ideas were taken up by his followers in England is attributed by C. I. Lewis to the difficulties of a logic of intension in which

---

[1] See [**63**, pp. 35–104].

A and B are equivalent if the class-concept of A is equivalent to the class-concept of B, rather than a logic of extension in which A is equivalent to B where the classes A and B consist of identical members. J. H. Lambert, the German mathematician, and Castillion, the French logician, working in 1803 experienced the same difficulty when they came to represent the inverse $A - B$; the problem being that if this means A BUT NOT B, i.e., the abstraction of B from A, then + and − are not true inverses. This can be seen in an example provided in [**115**]. If we have

$$\text{MAN} = \text{RATIONAL} + \text{ANIMAL}$$

then by Lambert's procedure we should also have

$$\text{RATIONAL} = \text{MAN} - \text{ANIMAL}$$

which is obviously false.

A number of innovations were made by Lambert in his calculus that anticipated Boole and De Morgan and Peirce. In particular, he used Greek letters such as α, ρ and ν to represent indeterminate quantities in the minor figures of the syllogism, as Boole was to use ν to represent the indeterminate quantity meaning SOME, ALL OR NONE. He also used fractional forms to represent the propositions of the syllogism, anticipating Boole's use of such forms in his calculus. Relations and their powers were also considered by Lambert, who introduced a notation for a relation that behaves like multiplication, thus anticipating the relative product of Peirce.

For example,

$$\text{if } f = \text{FIRE}, h = \text{HEAT and } a = \text{CAUSE}$$

then Lambert stated that[2]

$$f = a :: h$$

This equation can be translated as "fire is the cause of heat" where :: indicates that the relation a is applied to h.

On the other hand, Lewis opines that [**115**, p. 35]:

---

[2]See [**112**, p. 19].

> It is no accident that the English were so quickly successful after the
> initial interest was aroused; they habitually think of logical relations
> in extension. ... The record of symbolic logic on the continent is a
> record of failure, in England, a record of success.

As mentioned earlier in Section 2.3, an interest in symbolic procedures and abstract systems arose in England at the beginning of the nineteenth century when Continental mathematics using algebraic methods, developed by Laplace, Argobast and
Lagrange, influenced British mathematicians such as Babbage and Herschel to adopt
their ideas of symbolic notation, generalization and symmetry. However the period
1800–1830 also saw a revival of logic in England; the main protagonists being Thomas
Kirwan and Richard Whately. Mathematics and logic moved closer together as the revival of interest in the study and classification of the Aristotelian syllogism progressed.
Logic was now studied as a science by mathematicians, instead of being a subject regarded as an art form studied by philosophers. Common ground between logic and
mathematics included the study of language, concerns with generality, the use of symbols, analogy, symmetry and conciseness.

The main influences on Charles Peirce's algebraic logic came from the English
mathematicians Augustus De Morgan, George Boole and W. Stanley Jevons. However
a not inconsiderable influence was his own father Benjamin Peirce's work on linear
algebra. Let us now relate these influences to Charles Peirce's published works in the
area of algebraic logic, taken in strict chronological order.

3.2. The Logic of Augustus De Morgan and George Boole

3.2.1. Augustus De Morgan - Some Biographical Details. Born the fifth child
of Colonel De Morgan of the Indian Army, Augustus De Morgan spent the first seven
months of his life in Madura, India but lived in England thereafter. His father, who
died when he was ten, began the first forays into the education of his son (according
to a list of his early teachers drawn up by the young Augustus, in which he describes
himself as The Victim). His parents, who were strict evangelicals, also made sure that
Augustus attended church daily. What effect this had, can be gathered from the fact
that after his death Sophia, his wife sought out St. Michael's Church in Bristol, the site
of De Morgan's last school and found in the school pew [**41**, p. 8]:

...neatly marked on the oak wainscot partition, the first and second propositions of Euclid and one or two simple equations, with the initials A. De M. They were made in rows of small holes with the sharp point of a shoe-buckle.

His schooldays were not particularly happy ones apart from his obvious love of mathematics. He had lost his right eye from an illness acquired at birth in India and this ensured that he could never join in the games of his schoolfellows and also made him the recipient of many practical jokes which he was to abhor in later life. However he enjoyed Cambridge, graduating as fourth wrangler, where he was taught by Airy, Whewell and Peacock. But his religious scruples prevented him from pursuing his academic career here, as fellows were obliged to abide by the doctrines of the established Church of England. His wife later wrote, "Mr De Morgan never joined any religious sect, but I think he had most respect for the Unitarians."[3]

De Morgan's failure to graduate as Senior Wrangler at the Cambridge tripos greatly concerned and upset his mother. From this time he strongly disapproved of all competitive examinations. After an abortive attempt at law – his mother's choice when Augustus refused to consider holy orders – De Morgan was elected the first professor of mathematics at the new non-denominational London University at the young age of twenty-two in 1828. He was to remain here until 1866 when he resigned in protest at the decision of the College's council to refuse to appoint a well-known Unitarian to a vacant chair. He had previously resigned in 1831, again on a matter of principle but returned in 1836 when his successor died in an accident.

In 1837 De Morgan married Sophia Elizabeth Frend, daughter of William Frend, a neighbor who shared his religious beliefs. The young couple lived in Upper Gower Street in London. De Morgan loved the town and hated trees, fields and birds. He has been described as a man of great simplicity and vivacity of character, of affectionate disposition, and entire freedom from all sordid self-interest and having a voice of sonorous sweetness, a grand forehead, and a profile of classic beauty. However he was not a saint. De Morgan could be inflexible, "he held to his principles with a certain mathematical rigidity which excluded all possibility of compromise and gave ground

---

[3]Coincidentally Boole was also a Unitarian who felt himself unable to take up an academic post in England and later obtained the Chair in Mathematics at Queen's College, Cork, partly through the good offices of De Morgan. Charles Peirce was brought up in a Unitarian household but became an Episcopalian influenced by his first wife. His father-in-law was an Episcopalian minister.

for the charge of crotchetiness on some important occasions."[4] His wife was also to refer to him as uncompromising.

As well as his career as a university lecturer and as a part-time actuary, Augustus De Morgan wrote many influential textbooks and his work on double algebras was the forerunner of quaternions and contained the complete geometrical interpretation of -1. His work on triple algebras was an inspiration for Benjamin Peirce's *Linear Algebra*. Remembering her husband's pedagogical skills, Sophia De Morgan described how he loved his work and how his pupils were endeared to him by the interest they took in his teaching.[5]

He was remembered fondly by one of his students, Richard Hutton who claimed that in Mr. De Morgan's time, the mathematical classes of University College were quite as much classes in Logic as in Mathematics. De Morgan opposed competitive examinations and rote learning. He promised his students to set them papers which would ensure that cramming would be of absolutely no use. His method of assessing work was also original as he would judge the entire piece as a whole rather than use a marking scheme. Obviously a popular lecturer – famous students included Sylvester and Jevons – to make ends meet he took private students, of which one was Lady Ada Lovelace, Lord Byron's daughter. Lady Noel Byron was a very close friend of Sophia De Morgan. It was through this connection that Ada Lovelace was first introduced to Babbage's Analytical Engine, which she was to popularize later. However the story has an unhappy ending as Ada's later public dispute with her mother led to De Morgan and Babbage to take opposing sides in the quarrel, thus ending their close friendship.

Unlike Benjamin Peirce, De Morgan was not a committee man with two major exceptions. He was the first president of the London Mathematical Society, co-founded by his son George, and although he did not approve of joining societies, he was for many years Secretary of the Royal Astronomical Society. He wrote to John Herschel on October 16, 1832:

> My dear Sir John,
> ...I shall be very much obliged to you for all you have offered on
> the Catalogue, the Comet, and the Herscheliana. The crumbs which

---

[4]See [**193**].
[5]See [**41**].

fall from a rich man's table are good - astronomically, whatever they may be gastronomically.

This letter, apart from showing De Morgan's keen interest in astronomy, also shows his sense of fun and appreciation of a good pun. In another letter addressed to William Frend on September 1, 1834, he writes humorously of the reception of a friend of his, one Mr. Woolgar, by a "calculating machine maker" who can be none other than Babbage.

> I was very sorry to find when I came home that Mr. Woolgar had been very uncourteously received by B_____with my note. That unfortunate man will never rest until he succeeds in getting nobody's good word. He calculates very wrong (for a calculating machine maker) if he thinks such a thrower of stones as himself can stand alone in the world. It takes all his analysis and his machine to boot to induce me to say I will ever have any communication with him again.

This also shows De Morgan's mildy irascible nature.

His famous controversy with Sir William Hamilton, Professor of Logic and Metaphysics at Edinburgh, began in 1845, when he had already written his first paper "On the Syllogism: I" and had prepared *Formal Logic* for publication. He then asked Hamilton to send him some information on the history of the Aristotelian theory of the syllogism. Hamilton sent him some lecture notes which he had issued to his students as the requirements for a prize essay on logic, and a prospectus for a book on logic which as it turns out was never published. From 1839, Hamilton had been working on a logical system based on the principle of a quantified predicate as well as subject, thus yielding eight propositional forms of the syllogism instead of the usual four, enabling such propositions to be treated as statements of identity. De Morgan had been developing a numerically definite syllogism, which involved assigning values to the middle term of the syllogism and which co-incidentally also yielded eight forms. Hamilton claimed plagiarism and later that De Morgan had published Hamilton's idea of quantification under the impression that it was his own discovery. However De Morgan was not totally blameless in the argument, refusing to read two polite and conciliatory letters from Hamilton in 1847.

De Morgan saw himself and his great friend Boole as allies and proponents of the view that logic and mathematics had elements in common and could benefit from looking at such common properties. He wrote of Baynes' essay which had won Hamilton's prize in 1845: "I and Boole come in, without being named, for a lecture against meddling with logic by help of mathematics," and after Boole's death in 1864, he was to write, "Of late years, the two great branches of exact science, Mathematics and Logic, which had long been completely separated, have found a few common cultivators. Of these Dr. Boole has produced far the most striking results."

De Morgan respected Hamilton and called him affectionately an archsyllogist. The two disputants also seem to have enjoyed the controversy and De Morgan in particular was very fond of replying to any criticisms in print, thus fueling the debate. The resulting publicity delighted him as it had the effect of throwing his system of logic into greater prominence and he seems to have achieved the upper hand in the whole quarrel. Each new criticism from Hamilton or his followers, Baynes and Mansel, inspired him to greater efforts in his work. This public argument had inspired Boole to write *The Mathematical Analysis of Logic, Being an Essay towards a Calculus of Deductive Reasoning* [14] in 1847, which was published in the same year as De Morgan's first book, *Formal Logic: or The Calculus of Inference, Necessary and Probable* [33], and so in a sense modern logic owes a great debt to Sir William Hamilton.[6]

The death of De Morgan's son George in 1867 and his daughter Helen in 1870, "gave a fresh shock to his nerves and he afterwards sank gradually and died on 18 March 1871."[7]

3.2.2. De Morgan on the Syllogism - Introduction. De Morgan's work on the logic of relations was the first extensive attempt to record relational arguments. It has been called his greatest achievement in logic. Daniel Merrill in his book *Augustus De Morgan and the Logic of Relations* [124, p. viii], showed how the development of De Morgan's logic of relations arose from two strands. Firstly De Morgan's emphasis on the importance of relational inference in categorical propositions and secondly his attempt to express syllogistic logic within the framework of his logic of relations.

---

[6]These two books will henceforth be referred to simply as *Mathematical Analysis of Logic* and *Formal Logic* respectively.

[7]See [193].

The revival of an interest in logic in the early eighteenth century began with the publication in 1826 of Richard Whately's *Elements of Logic*. It was this book in the hands of his older brother Jem, that first introduced the young Charles Sanders Peirce to formal logic in 1851. Logic was presented as more akin to science than to art, with its own abstract structures and investigation into common systems of reasoning. In this work the formal and general aspects are emphasized with logic being firmly placed in the Aristotelian syllogism. The emphasis on the formal and general interested mathematicians such as De Morgan who then sought to extend logic beyond traditional syllogistic logic. Euclidean geometry was held to be 'the supreme example of demonstrative reasoning' ... 'using rigorous logic to generate innumerable truths from a few evident first principles.'[8]

However, attempts to syllogize Euclid were generally unsuccessful. It may be the case that De Morgan's early interest in Euclidean geometry led the way to the works of Aristotle, and the syllogism as a means of deductive reasoning. De Morgan had always been attracted to the rigor of mathematical reasoning and thought it an ideal way of training the young mind. His interest in particular in geometrical reasoning led him to consider logic. His first work in logic was titled, *First Notions in Logic, Preparatory to the Study of Geometry* [**32**].

Other logicians working in the field at the same time were Thomas Reid, Sir William Hamilton of Edinburgh, and Henry L. Mansel. Thomas Reid was critical of syllogistic logic as being insufficient to express all logical inference. In his "A Brief Account of Aristotle's Logic" which formed part of his *Works,* Vol. II, published in 1863 and annotated by Sir William Hamilton, he argued that the use of the syllogism was limited and not applicable to most mathematical reasoning, which involved relational propositions rather than categorical ones of the usual subject-predicate form, but instead consisted of two terms and a relation; an argument rejected by Hamilton.

In the third edition of his *Artis Logicae Rudimenta, from the Text of Aldrich* of 1856, Mansel, a follower of Hamilton, also looked at attempts to express Euclid's work syllogistically as well as noting that relational propositions do not fall within the traditional syllogism, as they are dependent on the relations used rather than simply the form of the propositions. De Morgan began by differing from both Hamilton and Reid by allowing IS EQUAL TO to be an alternative form of the copula IS. He considered that any transitive and symmetric relation would serve equally well as the traditional

---

[8]See [**124**, p. 11].

copula IS, the transitive and symmetric nature of the identity relation being the important factor in syllogistic reasoning. However he progressed to allowing any number of copulas in *Formal Logic*. This doctrine of the abstract copula accordingly provoked the controversy sketched above with traditional logicians who allowed only one copula IS.

3.2.3. "On the Syllogism: I" of 1847 to "On the Syllogism: IV" of 1860. We now examine De Morgan's system of logic by considering his series of four logical papers entitled "On the Syllogism" contributed to the *Transactions of the Cambridge Philosophical Society* between 1846 and 1860. In "On the Syllogism: I. On the Structure of the Syllogism," read on November 9, 1846, De Morgan set out to reform the traditional system of the syllogism. He stated:[9]

> ... the general impression among writers seems to be that there cannot exist any other theory of the syllogism except that derived from Aristotle. ... we need another which is self-consistent, true and comprehensive.

The main logical concepts introduced in this memoir are those of the *contrary* of a term and the *universe* of a proposition. The notion of x being not X, the contrary of X, is clarified in the following: "everything in the universe is either X or x."[10] The universe here stands for the universe of a proposition or a term, where the universe is strictly limited to the scope of that proposition or term, a concept which was developed in [**34**], as the universe of discourse.

He now introduced his own logical notation to represent the four traditional forms of the syllogism using a system of brackets and dots:

A	$X)Y$	EVERY X IS A Y
O	$X:Y$	SOME Xs ARE NOT Ys
E	$X.Y$	NO X IS Y
I	XY	SOME Xs ARE Ys

In an addition to the memoir dated 1847, De Morgan wrote:

---

[9]See [**90**, p. 1].
[10]See [**90**, p. 3].

> Since this paper was written, I found that the whole theory of the
> syllogism might be deduced from the consideration of propositions
> in a form in which *definite quantity* of assertion is given both to the
> subject and the predicate of a proposition ... From the prospectus of
> an intended work on logic, which Sir William Hamilton has recently
> issued, ... as well as from information conveyed to me by himself
> in general terms, I should suppose it will be found that I have been
> more or less anticipated in the view just alluded to.

In this De Morgan was mistaken, as his idea of the numerically definite syllogism does
not match up with the more general system of Hamilton. De Morgan introduced the
numerically definite syllogism later in this addition in several examples, e.g., EACH ONE
OF 50 Xs IS ONE OR OTHER OF 70 Ys. "On the Syllogism: I" was developed into Chapters
IV, V, VIII, and X, of *Formal Logic*.

At the start of his second memoir, "On the Syllogism: II. On the Symbols of
Logic, the Theory of the Syllogism and in particular of the Copula," read on February
25, 1850, De Morgan stated his aim of developing the syllogism

> ... with particular reference to the application of symbols, [to form]
> ... the algebra of the laws of thought.

The use of this phrase is significant as it was to be four years before Boole's *An Investigation of the Laws of Thought, on which are founded the mathematical theories of
Logic and Probabilities*[11] was published. He also showed that he had now realized that
there was little connection between his numerically definite syllogisms in which he
assigned a numerical value to subject or predicate, and the SOME OR ALL quantification
used by Hamilton. Boole's *Mathematical Analysis of Logic* had also been published
and De Morgan was careful to distance himself from the Boolean system. De Morgan
states:

> ... the methods of this paper have nothing in common with that of
> Professor Boole, whose mode of treating the forms of logic is most

---

[11] This work [15] will be referred to as simply *Laws of Thought* henceforth.

worthy the attention of all who can study that science mathematically, and is sure to occupy a prominent place in its ultimate system.[12]

De Morgan's own work on algebra was influenced by Peacock's espousal of the generality of algebra at Cambridge, but he had his doubts about his principle of the permanence of equivalent forms, instead preferring to justify his results by the truths of the conclusions that could be drawn from them. His work on functional equations also drew heavily on his algebraic tradition.[13] In "On the Syllogism: II" he made the analogy between solving an equation by elimination in algebra, and making an inference by describing "the object of thought in terms of others, by means of an assertion in which they are all involved" in logic.

However De Morgan was clear that algebra and logic do not always agree, just as there are some forms of inference that cannot be expressed syllogistically. He recalled his challenge made in *Formal Logic*, to deduce EVERY HEAD OF A MAN IS THE HEAD OF AN ANIMAL from EVERY MAN IS AN ANIMAL using traditional syllogistic logic. It is interesting to note that at this early stage, De Morgan is already considering the place of relations, e.g., HEAD OF _____ in syllogistic logic.[14]

New notation for the four traditional propositions was introduced, with greater use of brackets. The symbol) was used to signify the universal quantifier, EVERY, the symbol (signified the particular quantifier, SOME, and a dot or period signified negation.

A	$X))Y$	EVERY X IS A Y
O	$X(.(Y$	SOME Xs ARE NOT Ys
E	$X).(Y$	NO X IS Y
I	$X()Y$	SOME Xs ARE Ys

In De Morgan's system, obtaining inferences consist in erasing the symbols of the middle term, then the remaining symbols show the inference, e.g.,

$$()))   \quad \text{or} \quad   X()YY))Z$$

---

[12]See [**90**, p. 22].

[13]See for example [**133**, p. 201–252].

[14]See page 131 for the relational treatment of this syllogism by Peirce in 'Notes', MS 152:Nov.– Dec. 1868. See also [**74**, Section 2.4] for a more detailed account of De Morgan's analogies between logic and mathematics.

or

Some Xs are Ys, all Ys are Zs.

The conclusion is

$X()Z$, Some Xs are Zs $X()Y))Z$ gives $X()Z$

or

$())) $ gives $()$.

It is clear that his algebraic background has given him the inspiration to look anew at logical concepts. He wrote, "all my perception of complete quantification of both terms was derived from the algebraical form of numerical quantification."[15]

Another algebraic analogy which De Morgan adopted as seen earlier, was the idea of opposites, whether opposite algebraic terms or relations. So he was especially interested in logical contrary terms. On page 37, he outlined his rule for dealing with contraries as follows:[16]

> The rule of transformation is :- To use the contrary of a term with-out altering the import of the proposition, alter the curvature of its parenthesis, and annex or withdraw a negative point, e.g.,
>
> $$X(.)Y - -x))Y - -X((y - -x).(y.$$

This system also has the advantage that positive propositions contain an even number of dots (or none), while negative propositions contain an odd number, thus strengthening the algebraic analogy. In a later section entitled "On the Theory of the Copula and its Connexion with the Doctrine of Figure", on page 50, in introducing his theory of the abstract copula, De Morgan also revealed his strong algebraic basis:

> In my work on Formal Logic, I followed the hint given by algebra, and separated the essential from the accidental characteristics of the

---

[15]See [**90**, p. 35].

[16]Unfortunately at the end of this line, De Morgan uses a full stop, which somewhat confuses the issue.

copula, thereby shewing the conditions of invention for a copula different from the ordinary one.

De Morgan held that there are two necessary conditions for the copula, (which is now represented as a long dash so that $X \longrightarrow Y$ means SOME Xs ARE Ys) which make it sufficient for all forms of inference,

$$\text{Transitivity} \quad X \longrightarrow Y \longrightarrow Z = X \longrightarrow Z$$
$$\text{Symmetry} \quad X \longrightarrow Y = Y \longrightarrow X$$

The negative copula X - - Y (which had the interpretation SOME Xs ARE NOT Ys) was also introduced on page 51, the definition that within the universe of the syllogism, either $X \longrightarrow Y$ or X- -Y must be true. When contrary terms are introduced, the copular condition further required is that either $X \longrightarrow Y$ or $X \longrightarrow y$ should hold for any X.

On page 56, De Morgan used analogy with algebraic equations to show that inference in logic can be expressed in terms of composition of relations:

> The deduction of $y = \phi\psi z$ from $y = \phi x$, $x = \psi z$ is the formation of the composite copula $= \phi\psi$. And thus may be seen the analogy by which the instrumental part of inference may be described as the elimination of a term by composition of relations.

This was the key idea of De Morgan's theory. He showed that the copula is could be replaced with a more general relation that was both transitive and convertible and that all inference in logic could now be represented by composition of such relations. Finally on page 63, he introduced for the first time subscript and superscript notation, the subscript prime to stand for the universal quantifier and the superscript prime to stand for the existential quantifier ONE OR MORE. He writes:

> X, $))'Y$ may stand for EVERY X GIVES TO ONE OR MORE Ys.

In "On the Syllogism: III" [37], read on February 8, 1858, De Morgan had the opportunity to address again the relation between logic and mathematics that he briefly introduced in "On the Syllogism: II". He writes on page 77:

> The separation of mathematics and logic which has gradually arrived
> in modern times, has been accompanied, as separations between near
> relations generally are, with a good deal of adverse feeling. Great
> names in each have written and spoken contemptuously of the other;
> while those who have attended to both are aware that they have a
> joint as well as a separate value.

De Morgan made a plea for greater understanding from both logicians and mathematicians and prophetically in a footnote on page 78 he wrote of a new discipline:
mathematical logic.

> As joint attention to logic and mathematics increases, a logic will
> grow up among the mathematicians, distinguished from the logic of
> the logicians by having the mathematical element properly subordi-
> nated to the rest. This mathematical logic ... will commend itself
> to the educated world by showing an actual representation of their
> form of thought.

In "On the Syllogism: IV" [**40**], read on April 23, 1860, traditional syllogistic
logic is for the first time expanded to show the relations between the subject and predicate. Where the copula is was deemed sufficient for the traditional Aristotelian approach, De Morgan's theory of the abstract copula demanded a theory for expressing
such relations. However not much more than the introduction of the notation and the
development of tables of syllogisms and the various combinations of relations, is attempted by De Morgan here. The notation for the theory of relations is introduced on
page 215 in the following way:

$$X..LY \quad \text{means} \quad X \text{ is } L \text{ of } Y$$
$$X.LY \quad \text{means} \quad X \text{ is not an } L \text{ of } Y$$

This notation seems to be a combination of the dot notation signifying negation
in De Morgan's earlier syllogistic logic, (and therefore two dots for affirmation), and
functional notation $y = \varphi x$. Mathematical functional symbols are not used perhaps
because of the general objection to the use of mathematical symbols in logic that was
prevalent at this time, as we have seen earlier in the De Morgan-Hamilton dispute. On
page 221, we have the first introduction of composition of relations:

$$X \,..\, L(MY) \quad \text{means} \quad X \text{ is one of the } L\text{s of one of the } M\text{s of } Y$$

He was clear that $L(MY)$ is equivalent to both $(LM)Y$ and LMY.

Although De Morgan realized that composition of relations was equivalent to logical multiplication or comprehension of classes, he did not consider logical addition, also known as aggregation or extension, of relations apart from stating the definition, i.e., $X\,..\,(L-M)Y$. Here X is either one of the Ls of Y or one of the Ms of Y, or both. He introduced the universal quantifier EVERY by using the superscript prime notation that appeared in "On the Syllogism: II" with the only difference being that the prime was used as a subscript in the earlier paper rather than a superscript.

$$LM' \quad \text{means} \quad \text{an } L \text{ of every } M$$
$$L-M \quad \text{means} \quad \text{an } L \text{ of none but } M\text{s}$$

This superscript and subscript notation first used by De Morgan in "On the Syllogism: II" to express quantification in syllogistic logic and here in the logic of relations, appeared again in Peirce when he came to express quantification in *Logic of Relatives*.

On page 222, the converse relation L^{-1} and the contrary relation l are defined in the following way:

$$Y\,..\,L^{-1}X \stackrel{\text{def}}{=} X\,..\,LY$$
$$X\,..\,lY \stackrel{\text{def}}{=} X.LY.$$

so that

$$X\,..\,L^{-1}Y \quad \text{means} \quad \text{If } X \text{ is the master of } Y, \text{ then the converse}$$
$$\text{relation would be } Y \text{ is the servant of } X$$
$$X\,..\,lY \quad \text{means} \quad X \text{ is some not-}L \text{ of } Y$$

De Morgan then proceeded to sketch rough proofs of various theorems of converses and contraries, e.g., the contrary of a converse is the converse of the contrary.

$$X\,..\,LY \Rightarrow Y\,..\,L^{-1}X \Rightarrow Y.\text{not-}L^{-1}X$$

but also

$$X\,..\,LY \Rightarrow X.\text{not-}LX \Rightarrow Y.(\text{not-}L)^{-1}X$$

so

$$\text{NOT-}L^{-1} \equiv (\text{NOT-}L)^{-1}.$$

Similarly he showed that the conversion of a compound relation converts both components and inverts their order, i.e.,

$$(LM)^{-1} \text{ is } M^{-1}L^{-1}$$

On page 224, Theorem K is introduced:

$$LM))N \Rightarrow L^{-1}n))m \ \& \ nM^{-1}))l.$$

The pattern here is to take the converse of one relation and then interchange the contraries of the other two relations.

He defined identical relations and used the symbol || for such equivalent relations. It is clear that when used with relations, $L||M$ is equivalent to $L))M$ and $M))L$. De Morgan also defined convertible relations, i.e., relations that are their own converses. For example, $X..LY$ gives $Y..LX$. This is our modern notion of symmetry. He pointed out that LL^{-1} is convertible and comes very close to defining the identify relation. We have on page 226:

> Take identity, for example: it is the very notion of identity between X and Y that $X..LL^{-1}Y$ for every possible relation L in which X can stand to any third notion.

Transitive relations are defined as those where a relation of a relation is a relation of the same kind as symbolized by: $LL))L$, $LLL))LL))L$, etc.

In summary, it is clear that the revolutionary approach to traditional logic De Morgan has developed in this paper lies in the replacement of the syllogism by "a composition of two relations into one," as he writes on page 238. The remainder of the paper is taken up with a comparison between logic and mathematics (i. e. algebra), which we will now consider in greater detail.

3.2.4. De Morgan's Thoughts on the Relations between Logic and Mathematics.
Throughout "On the Syllogism: IV" De Morgan repeatedly stressed the close relationship between logic and algebra. He thought that algebra was the most natural

form of expressing logic and that it was of benefit to compare the similarities between the two disciplines. On page 235 he wrote:

> There is no more limit to the formulae of thought than to the formulae of algebra ... There is identity or difference in every possible logical judgement: there is equation or inequation in every possible algebraical judgement.

He also questioned the orthodox logicians who opposed him because he wanted to reform the system and not adhere to its standard forms. He wrote humorously on page 239:

> Nothing that I know of can be written all in syllogism, except mathematics: and this is merely because, out of mathematics, nearly all the writing is spent in loading the syllogism, and very little in firing it.

As a mathematician, De Morgan felt it natural to apply mathematics to logic, and yet he was well aware of the opposition from traditional logicians. Towards the end of the memoir on page 241 he wrote:

> It is to algebra that we must look for the most habitual use of logical forms. ... Not that I by any means take it for granted that all those who have cultivated both sciences will agree with me.

The resurgence of interest in logic that had arisen in the early part of the nineteenth century, inspired by the works of Whately, Thomson, Hamilton etc. was bound to have an influence. In fact De Morgan considered Archbishop Whately to be "the restorer of logical study in England.". Aristotelian syllogisms were found wanting and in need of improvement. De Morgan was part of the movement that applied algebraic methods and techniques to effect such reforms in developing the new mathematical logic.

Although logic could be defined broadly as the inquiry into truth and falsehood and narrowly as the investigation of the Aristotelian syllogism, De Morgan favored Kant's definition as "the science of the necessary laws of thought," a definition that echoed both Benjamin Peirce's definition of mathematics and Boole's *Laws of Thought*. For De Morgan, mathematics is not part of logic; the main distinction being that logic deals with the pure *form* of thought without any considerations of *matter*. Like Boole, he felt that logic and human reasoning could be symbolized in the forms of algebra, although he left this largely to Boole, contenting himself with reforming the traditional Aristotelian syllogism and developing a theory of relations.

In particular, he considered the similarities between logic and mathematics, and noted that it was the similarity between the opposite relations of + and - in algebra and the many opponent notions in logic, e.g., affirmative and negative, existent and non-existent etc., that led him to further consider this area; in particular whether these opponent notions can be interchanged. He wrote on page 23 of "On the Syllogism: II":

> The suggestions of symbolic notation have led me to more recogni-
> tion than is usually made of harmonies which exist among various
> pairs of opponent notions common in logical thought.

and later on page 26:

> I think it reasonably probable that the advance of symbolic logic
> will lead to a calculus of opposite relations, for mere inference, as
> general as that of + and - in algebra.

Looking more closely at the parallels between algebra and logic he drew the following similarities represented in tabular form in 3.1:

LOGIC	ALGEBRA
inference	elimination
assertions	equations
middle terms	eliminated quantities

TABLE 3.1. Similarities between Logic and Algebra

He made the analogy between solving an equation by elimination in algebra, and making an inference by describing "the object of thought in terms of others, by means of an assertion in which they are all involved" in logic. However he asserted that algebra and logic do not always agree, just as there are some forms of inference that cannot be expressed syllogistically e.g. his challenge of deducing EVERY HEAD OF A MAN IS THE HEAD OF AN ANIMAL syllogistically from EVERY MAN IS AN ANIMAL. As we have seen, all of De Morgan's innovations such as quantification of subject and predicate by a numerically definite amount, his development of opposite or contrary terms or relations, his replacement of the copula IS by an abstract relation and finally his use of composition of relations to represent logical inference, were inspired by algebra. Even though he did not use mathematical methods or notation in an attempt to make his work accessible to non-mathematical logicians, many of his basic concepts listed above were derived from algebra. He believed that the growth of logic had been stunted by its separation from mathematics, and wrote prophetically in [**90**, p. 345]:

> I believe, and I am joined by many reflecting persons, among students both of logic and of mathematics, that as the increasing number of those who attend to both becomes larger and larger still, a serious discussion will arise upon the connection of the two great branches of exact science. ... The severance which has been widening ever since physical philosophy discovered how to make mathematics her own especial instrument will be examined, and the history of it will be written.

Peirce's next paper on algebraic logic was inspired by "On the Syllogism: IV." His initial works, "Harvard Lecture III" [**142**] and "Harvard Lecture VI" [**143**] were written as a follower of Boole and his later papers such as "On an Improvement in Boole's Calculus of Logic" [**144**] and 'Upon the Logic of Mathematics' [**146**] extended and improved the Boolean calculus along the lines already suggested by Jevons in his *Pure Logic* [**103**], as far as the operation of logical addition between classes was concerned. Murphey writes:

> The discovery of the calculus of relations was one of the most important events in Peirce's philosophic career. For although Peirce's contributions to logic were many and varied, it is primarily upon his work in relations that his fame as a logician is based. To De Morgan belongs the credit for originating modern relation theory, but it was Peirce who developed it, and virtually all of the calculus of relations of the Boole-Schröder algebra was his creation. Not until the *Principia* appeared was Peirce's work superseded and then only by a theory based in large part upon his own. [**129**, p. 152]

For Peirce now attempted to include De Morgan's work on the logic of relations with Boolean algebraic logic in the following work, the importance of which was not lost on Peirce himself. In his Lowell lectures of 1903, he wrote:

> In 1870 I made a contribution to this subject [logic] which nobody who masters the subject can deny was the most important excepting Boole's original work that ever has been made.

Other mathematicians agree with Peirce in this respect, namely the British mathematicians W. K. Clifford in 1877 who said:

> Charles Peirce . . . is the greatest living logician, and the second man since Aristotle who has added to the subject something material, the other man being George Boole, author of *The Laws of Thought*.[17]

Daniel Merrill has claimed that the 1870 memoir is "one of the most important works in the history of modern logic."[18] Robert Burch argues that "Peirce's 1870 work contains a logic of relations at least as powerful in expressive capability as first order predicate logic with identity."[19]

3.2.5. Features of George Boole's Logic. Peirce had the greatest respect for George Boole. He referred to him as "a man who united a genius for mathematics with a high originality as a logician."[20] Boole was born in 1815 and began his study of mathematics at the age of sixteen, starting off with the works of the French mathematicians Lacroix, Laplace and Lagrange. In 1833 he became the headmaster of a school near his home town of Lincoln. Five years later he was corresponding with E. F. Bromhead, a former member of the Analytical Society of Cambridge, Duncan Gregory and Robert Murphy on matters algebraic, thus reinforcing the effect of his earlier studies of Lagrangian techniques, i.e. of reducing physical problems to purely algebraic terms.

---

[17]From a letter of Youman reporting a visit with Clifford. See [**61**, p. 340].

[18]See [**56**, p. xlii].

[19]See [**21**, p. 206].

[20]See [**55**, p. 404].

In particular his friendship with David Gregory provided Boole with the extra stimulus he needed for the study of symbolic algebra and its application to the calculus of operations. Augustus De Morgan's friendship beginning in 1842 was also valued by Boole although these two men exerted little mathematical or logical influence on one another. However De Morgan did influence Boole's work if only indirectly in one very important respect, namely, that the dispute between Hamilton and De Morgan over the discovery of the quantification of the predicate led Boole to write his *Mathematical Analysis of Logic*.

The famous dispute between De Morgan and Sir William Hamilton of Edinburgh over the validity of the study of mathematics and whether Hamilton or De Morgan had precedence in the discovery of the quantification of the predicate, led Boole to compose *Mathematical Analysis of Logic* in 1847. The analyst Charles Graves saw the manuscript of this work prior to its publication and made some modifications incorporated by Boole.[21]

In 1849 Boole was elected Professor of Mathematics at Queen's College, Cork. Here he was able to develop his work in *Mathematical Analysis of Logic*, philosophically as well as mathematically. The fruit of his labor was *Laws of Thought*. Up to the early 1850s Boole had used symbolic methods in applications in analysis, e.g., using symbolic notation such as the letter D for the differential operator $\frac{\partial}{\partial x}$ in formulating a general method for the solution of certain differential equations with variable coefficients.

Drawing on the analogy between algebraic equations and syllogistic logic, Boole laid down the principles that were to form his general method in logic. The basis of his algebraic logic as set forth in *Mathematical Analysis of Logic* and *Laws of Thought* consisted of a three stage procedure:

  i) the formulation of problems of logic in terms of equations,
 ii) the solution of these equations
iii) the interpretation of the results obtained.

Boole tried to widen the scope of traditional syllogistic logic which had a restricted area of application and in *Mathematical Analysis of Logic* and especially in *Laws of*

---

[21]See [**133**, pp. 494–499].

Thought, pointed out instances which could not be treated by Aristotelian logic. On page 33 of *Mathematical Analysis of Logic* we have:

> The Aristotelian canons, however, besides restricting the order of the terms of a conclusion, limit their nature also.

He argued that scholastic logic, which until that time consisted of set patterns of Aristotelian syllogisms classified by letters such *A, E, I, O, U* or names such as *Barbara, Celarent*, etc., was not a proper foundation for logic. On page 10 of *Laws of Thought* he wrote:

> ...syllogism, conversion, etc., are not the ultimate process of Logic. It will be shown in this treatise that they are founded upon, and are resolvable into, ulterior and more simple processes which constitute the real elements of method in Logic.

Boole saw that algebra and logic had a special relationship. In both *Mathematical Analysis of Logic* and *Laws of Thought* he showed that problems of logic were easily expressible in the form of algebraic equations with symbols that varied from algebraic symbols only in interpretation. In fact algebra and logic were to George Boole two branches of a wider science.

Furthermore he felt that algebra, logic and even language were all expressions of a general calculus of symbols, a philosophical language that would help to interpret and therefore understand the workings of the mind. It is interesting to note at this point that although the young Charles Peirce was to eagerly embrace George Boole's algebraic logic as a means of resolving problems of logic through algebraic equations, he had little sympathy for the philosophical position of Boole.

Rather as Benjamin Peirce defined mathematics, Boole defined logic as a philosophy from which all the deductive sciences are developed. Although in both *Mathematical Analysis of Logic* and *Laws of Thought* he had successfully expressed logical ideas in mathematical form and logical reasoning by mathematical processes, to Boole, logic was not just applied mathematics. Grattan-Guinness makes this clear in [**69**]:

> ...neither subject is an application of the other but each is a particular case of the universal calculus.

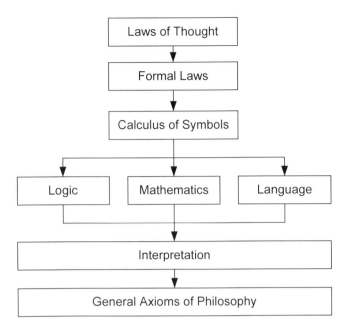

FIGURE 3.1.   Laws of Thought

In summary, George Boole sought to express logic, which had up to this time, been commonly expressed by Aristotelian syllogisms, in the form of mathematical and in particular algebraic language. Hailperin [**78**] has shown that Boole's system can be interpreted as an axiom system for signed heaps.

3.2.6. Divergences between Algebra and Logic in Boole. Boole was also a pioneer in establishing the close relationship that was to exist between algebra and logic which had until that time had been seen as a science and an art respectively.[22] As we have shown earlier he considered algebra and logic to be separate branches of a universal language of symbols. This is not to say that he did not fully realize the differences between algebra and logic. He highlighted in *Mathematical Analysis of Logic* and *Laws of Thought* three main areas of divergence:

---

[22]See [**75**, pp. xxviii–xxxv] for a discussion of Boole's quest for the foundations of his logic.

1) Division
2) The Index Law
3) Interpretation of Symbols.

3.2.6.1. *The operation of division is omitted from the logic of Boole.* Boole recognized the fact that the algebraic process of division and its use in eliminating unknown variables in solving algebraic equations had no logical counterpart:

> ...it cannot be inferred from the equation $zx = zy$ that the equation $x = y$ is also true. In other words, the axiom of algebraists, that both sides of an equation may be divided by the same quantity, has no formal equivalent here.[23]

The absence of a logical counterpart to the algebraic operation of division in both *Mathematical Analysis of Logic* and *Laws of Thought* was one of the first areas of improvement for Peirce.[24] At page 100 in *Laws of Thought* Boole also pointed out the difference between algebra and logic when eliminating variables. He was able to eliminate an indefinite number of variables in logic using the Index Law instead of only a finite number depending on the number of equations as in algebra:

> At present I wish to direct attention to an important, but hitherto unnoticed, point of difference between the system of Logic, as expressed by symbols, and that of common algebra, with reference to the subject of elimination. In the algebraic system we are able to eliminate one symbol from two equations, ...$n - 1$ symbols from n equations...
>
> But it is otherwise with the system of logic ...From a single equation an indefinite number of such symbols may be eliminated.

3.2.6.2. *The Index Law holds for all logical symbols, thus restricting logic to an algebra taking only the numerical values of 0 and 1.* In algebra there is no restriction in the arithmetic value of the variables, only in the number of solutions. However in Boole's calculus of logic, the symbols are restricted to the numerical values of 0 and 1. This is a consequence of the fact that for Boolean algebra every symbol obeys the

---

[23]See [**15**, p. 36].

[24]See [**133**, pp. 562–569] for instances of implicit division in Boole.

Index Law $x^2 = x$, sometimes written as the Duality Law $x(1 - x) = 0$.[25] At page 38 in *Laws of Thought* Boole writes:

> We have seen that the symbols of logic are subject to the special law, $x^2 = x$. Now of the symbols of Number there are but two, viz. 0 and 1, which are subject to the same formal law ... Hence, instead of determining the measure of formal agreement of the symbols of Logic with those of Number generally, it is more immediately suggested to us to compare them with symbols of quantity *admitting only of the values 0 and 1* ... The laws, the axioms, and the processes, of such an Algebra will be identical in their whole extent with the laws, the axioms, and the processes of an Algebra of Logic ... Upon this principle the method of the following work is established.

3.2.6.3. *Interpretation of the symbols in Boole's logic.* In [**15**, p. 6] Boole states:

> ... any agreement which may be established between the laws of the symbols of Logic and those of Algebra can but issue in an agreement of processes. The two provinces of interpretation remain apart and independent, each subject to its own laws and conditions.

Whereas algebraic symbols in arithmetic represent numerical quantities, logical symbols as used by Boole represent objects, qualities, propositions, mental operations of selection or even instances of time. The meaning of these symbols is quite distinct, depending on which text, either *Mathematical Analysis of Logic* or *Laws of Thought* is followed. On the one hand they stand for operations, i.e., mental process of selecting objects or objects with qualities as is the case in *Mathematical Analysis of Logic*, or on the other hand for the objects or qualities themselves as in *Laws of Thought*.

He also had alternative meanings for these symbols when he divided his logical expressions into primary and secondary propositions. Boole restricted his use of language to nouns, adjectives and prepositions:[26]

a) nouns specify classes within the Universe, such as MEN within HUMANS

---

[25]For the possible parentage of this law in Leibniz see [**75**, p. xliii].

[26]See [**75**, p. xxxiii].

b) adjectives determine sub-classes, such as GOOD MEN within MEN

c) prepositions expressed the connectives EXCEPT (-), exclusive OR (+) and AND (.)

Primary propositions were propositions of the form X IS Y and secondary propositions were propositions about propositions, i.e., of the form IF A IS B THEN C IS D.[27] In *Laws of Thought* when considering whether propositions are true, such propositions being secondary propositions, he interpreted his symbols as the periods of time for which such propositions are true. This philosophical notion of time was to cause later mathematicians some difficulties, and he was careful to ensure that those who did not agree with his idea could nevertheless still successfully use his method. He writes at [**15**, p. 164]:

> ... when those laws and those forms are once determined, this notion of time (essential as I believe it to be ...) may practically be dispensed with.

The use of algebraic symbols by Boole will be considered in greater detail later in this section.

Having established this close relationship between algebra and logic, he was to try a different approach and move away from his key idea as stated at [**15**, p. 12]:

> Logic – its ultimate forms and processes are mathematical.

This is shown in his subsequent logical work, in which he introduced a philosophic approach to logic using ordinary language and alternative symbols instead of mathematical ones. In this period from 1855 – 1860 he produced a series of manuscripts for a book on the philosophy of logic on the nature of logic, reasoning and the use of symbolism as distinct from mathematical notation in logic. However this venture proved to be not entirely successful. The more the use of such mathematical symbols was avoided, the more attention was drawn to their absence. Another disadvantage which explained why no later logicians were to follow along these lines, was that any attempt

---

[27]In Boole's propositional logic, the main distinction drawn is that between *categorical* propositions such as propositions about things – SOME X IS Y – and *hypothetical* propositions such as propositions about propositions – IF X IS Y THEN Y IS Z. This is the terminology used in *Mathematical Analysis of Logic*, whereas in *Laws of Thought* he uses primary vs. secondary for categorical vs. hypothetical.

at reasoning with logical propositions stated independently of mathematics, involved convoluted and drawn out arguments. These philosophical reasonings of Boole were never published in his lifetime in part due to ill health towards the end of his life [**169**, p. 10].[28]

After having successfully produced several works including *Mathematical Analysis of Logic* and *Laws of Thought* on the method by which logic could be expressed in a purely mathematical way, which were highly regarded by contemporary mathematicians, why did Boole in his later years move away from this approach and try instead to express logic in a more philosophical way? One reason for his efforts was that he hoped to procure a wider audience for his work than he had previously obtained, and this could only be achieved if he dispensed with the mathematical notation. The *Athenaeum* for December 17th, 1864, published a brief and cutting obituary of Boole stating:

> The Professor's principal works were *An Investigation into the Laws of Thought*, and *Differential Equations*, books which sought a very limited audience, and we believe, found it.

Another possible reason is that Boole was afraid that scholars would rely too heavily upon the purely mechanistic and formal methods of his algebraic equations without recourse to what he called "their higher intellect" which he saw as incorporated in his philosophical approach.[29] This of course had found expression earlier in the famous Hamilton-De Morgan controversy when Hamilton, the philosopher, was able to claim the moral high ground over De Morgan, the mathematician.[30]

3.3. Peirce's Early Stages in Logic, 1865–1870

3.3.1. The Early Years. Charles Sanders Peirce was born in Cambridge, Massachusetts, in 1839, the second son of Benjamin Peirce. The influence of his famous father was strongly felt in Charles's academic development. Benjamin trained his son's concentration by means of rapid games of double dummy which he played with his son from ten in the evening until sunrise, sharply criticizing every error. Benjamin also had

---

[28]See also [**75**] for a discussion of these later manuscripts.

[29]See [**169**, p. 10].

[30]See [**133**, pp. 578–584] for Boole's symbolic logic as developed in his later years.

his own original pedagogic ideas. Instead of disclosing general principles or theorems to his son, Benjamin would present him with problems, tables or examples and encourage him to work out the principles for himself.

Charles' introduction to logic came from Whately's *Elements of Logic*, a book belonging to his elder brother James that he saw by chance within a week or two of his twelfth birthday and immediately absorbed over a period of several days. However until this time, he had shown a decided preference for chemistry.[31] Charles' aunt and uncle, who had together translated from the German the standard American school textbook on chemistry, helped him to set up a chemical laboratory at home. In 1850, Charles at the age of eleven even wrote a "History of Chemistry". Charles was later to inherit his uncle's chemical and medical library shortly before entering Harvard College in 1855. It is easy to understand Charles' claim that he was brought up in a laboratory. His chemical career continued, when in the latter half of 1860, Charles was for six months a private student of Louis Agassiz, the famous biologist and close friend of Benjamin Peirce, to learn the Agassiz method of classification before entering the Lawrence Scientific School of Harvard University in the spring of 1861.

Although Charles' undergraduate scholastic record was poor, he was one of the youngest in his class to graduate. He was seventy-first out of ninety-one, …"apparently too young and of too independent a mind to distinguish himself under the rigid Harvard system of those days."[32] He did better two and a half years later when he graduated as a summa cum laude Bachelor of Science in Chemistry from the Lawrence Scientific School.

In July 1861 Charles entered the United States Coastal Survey of which his father Benjamin was the Superintendent. This would enable him to earn enough money to continue his studies in chemistry and in fact his first professional publication, in 1863 at the age of 23, was on "The Chemical Theory of Interpretation," [**156**]. He was to remain at the Coastal Survey as a geodeter, for the next thirty-one and a half years. His scientific work for the Survey naturally led him to the fields of astronomy, metrology, spectroscopy and geodesy especially pendulum research. On leaving the Survey he set up in private practice as a chemical engineer at the end of 1891.[33]

---

[31] See [**133**, pp. 654–667] for other links between chemistry and mathematics in the work of important mathematicians.

[32] See [**119**, p. 399].

[33] See [**55**, pp. xviii–xxi].

However if Charles Peirce saw his profession either as a chemist or geodeter, it is clear that at heart he was a logician. From the moment that he opened Whately's book he found it impossible to think of anything, including even chemistry, except as an exercise in logic. Charles initially saw logic as a classificatory science like chemistry. One of his earliest published papers in logic was called "On a Natural Classification of Arguments" [145].

The categories of Kant and Hegel also had a profound influence on his philosophy. He later broadened his concept of logic to include semiotics and all of his philosophy including his work on pragmatism[34] fell within the scope of his logic. Charles embedded himself in logic and came to believe that he made significant contributions to knowledge in this area. That said, there is no doubt that he saw the lighter side of logic too. In "Lowell Lecture VI" [55, p. 440] he gave five practical maxims of logic of which the first and last were "Beware of a syllogism." and "Everything can be explained." respectively.

In 1865 and 1866 Charles Peirce gave a series of lectures on the logic of science. These were intended primarily for graduates of Harvard University and each lecturer was expected to devote his lectures to the field and topics of his greatest competence, or on those which were at the forefront of current research [55, p. xxii]. In his Harvard Lectures III and VI, Peirce's topic was the algebraic logic of George Boole. I shall now analyze these Harvard Lectures, comparing them with *Mathematical Analysis of Logic* and *Laws of Thought*. It is quite evident that Peirce was to base his understanding of algebraic logic through a study of the former rather than the latter.

3.3.2. Peirce's "Harvard Lecture III" (1865). Charles Peirce claimed that he was first attracted to Boole's calculus of logic through the connection with probability. In Lecture III [142], he called algebraic logic "this curious branch of mathematics" and continued:

---

[34]Charles Peirce has been called the father of pragmatism. His famous definition of this philosophy, first published in Jan 1878 as an article, "How to Make Our Ideas Clear," in *Popular Science Monthly*, is:

Consider what effects, which might conceivably have practical bearings, we conceive the object of our conception to have. Then, our conception of these effects is the whole of our conception of the object.

> ...the knowledge of it enables us to solve readily all simple ques-
> tions of probability and to understand the general principles of solu-
> tion of the most difficult ones.

However Peirce was soon analyzing Boole's algebraic logic for its own sake.[35]

The follow analysis of Peirce's Harvard Lectures will consist of firstly considering the key ideas of logical symbols, multiplication, addition, equality, zero, unity,
numerical values, 0/0, subtraction, division and elimination in Boole's texts beginning
with *Mathematical Analysis of Logic* and then *Laws of Thought*. In each case we shall
then follow this with Charles Peirce's own interpretation of Boole, highlighting the
similarities and differences.

This analysis takes into account the fact that Boole divided his logic into two
parts. First, he considered primary propositions or a part/whole class logic he called
the logic of class, e.g., Some X is Y. Boole had called these categorical propositions
in *Mathematical Analysis of Logic*. He then introduced secondary propositions which
involved primary propositions united by a copula or conjunction, called hypothetical
propositions in *Mathematical Analysis of Logic*, e.g., If A is B then C is Y. Additional
complexity occurs because Boole was to alter his definitions of algebraic symbols for
his logic of secondary propositions.

A fundamental concept for Peirce in his "Harvard Lecture III" was Boole's use of
logical symbols. In *Mathematical Analysis of Logic* the logical symbol x stands for a
mental process rather than an entity, that is, an elective operation in which members of
a given class X are mentally selected from any given subject. However x can also be
used to express a class rather than an operation in the sense that $x = x(1)$.

On pages 15 and 16 of *Mathematical Analysis of Logic* we find:

> When no subject is expressed we shall suppose 1 (the Universe) to
> be the subject understood, so that we shall have

$$x = x(1)$$

---

[35]Boole's work on probability was secondary to his quest for a general method in logic, in spite of
Venn's statement to the contrary: that it was in fact largely for the purpose of improving the calculus of
probabilities that Boole devised his system. Instead the investigation of the theory of probability was to
Boole an application of his general method of algebraic logic.

the meaning of either term being the selection from the Universe of
all the x's which it contains, and the result of the operation being in
common language, the class X, i.e., the class of which each member
is an X.

Boole later amended this when he considered secondary propositional logic. On
page 49 of *Mathematical Analysis of Logic* he wrote:

The elective symbol x attached to any subject ... shall select those
cases in which the Proposition X is true.

It is clear that x is still to be considered as an operation of mental selection. But
in *Laws of Thought* a different approach is used. Here x as a logical symbol clearly
represents a class rather than an operation. By a class Boole meant a collection of
individuals or objects to which a particular name or description could be applied. He
defined x on page 28 of *Laws of Thought*:

Let us then agree to represent the class of individuals to which a
particular name or description is applicable, by a single letter, as x,
... let x represent ALL MEN, or the class MEN.

Later in his secondary propositional logic Boole went on to extend this definition
to include the period of time for which a proposition is true. On page 165 of *Laws of
Thought*, he writes:

Let x represent an act of the mind by which we fix our regard upon
that portion of time for which the proposition X is true; and let this
meaning be understood when it is asserted that x denotes the time
for which the proposition X is true.

Peirce followed *Laws of Thought* in defining his algebraic symbols as classes rather
than as operations of mental selection from classes and began on page 190 of "Harvard
Lecture III" by defining variables as classes of objects or qualities, e.g.:

h stands for the class of HORSES
b stands for the class of ALL BLACK THINGS.

Turning to the operations of multiplication, addition, subtraction, negation or class complementation and the relation of equality, under the chapter heading "First Principles"' on page 16 of *Mathematical Analysis of Logic*, Boole gave the following definition of multiplication:

> ...the product *xy* will represent, in succession, the selection of the class *Y*, and the selection from the class *Y* of such individuals of the class *X* as are contained in it, the result being the class whose members are both *X*s and *Y*s.

Notice the emphasis on the product *xy* as an operation of mental selection. This contrasts with the following definition is taken from *Laws of Thought*, page 28, where *xy* simply represents a class:

> *xy* shall be represented [by] that class of things to which the names or descriptions represented by *x* and *y* are simultaneously applicable.

Peirce, in line with *Laws of Thought* has:

> *bh* stands for the class of ALL BLACK HORSES.

Peirce made the point on page 190, that multiplication as in 3×2 implies three collections consisting of two individuals each gives six individuals. In the same way BLACK HORSES implies ALL BLACK THINGS each of which is a HORSE.

So it is clear that multiplication in Boole's algebraic logic is represented by intersection in a part/whole class logic not our modern-day intersection of sets in the Cantorian sense with $\cap$ but where class represents a collection of objects.

Addition is defined on page 17 of *Mathematical Analysis of Logic* very simply as aggregation of classes:

> $u + v$ representing the undivided subject, and *u* and *v* the component parts of it.

However a serious omission in *Mathematical Analysis of Logic* to be corrected in *Laws of Thought* is that there is no mention that the classes must be disjoint. This operation of conjunction also appears in *Laws of Thought* but in addition it is made quite clear in *Laws of Thought* on page 33 by using an analogy with language, that *b* and *h* have no common members, i.e., that the classes are quite distinct.

> In strictness, the words AND, OR, interposed between the terms descriptive of two or more classes of objects, imply that those classes are quite distinct, so that no member of one is found in the other. In this and in all other respects the words AND, OR are analogous with the sign + in algebra, and their laws are identical.

Charles Peirce himself defined addition as the operation of aggregation or conjunction used both in *Mathematical Analysis of Logic* and *Laws of Thought*, e.g. *h* + *c* represents the class of ALL HORSES AND COWS. But he made clear in this definition that the operation of addition is only applicable to disjoint classes, thus showing that *Laws of Thought* rather than *Mathematical Analysis of Logic* was his main text, since as shown above, these classes are only explicitly disjoint in *Laws of Thought*. In his "Harvard Lecture III," Peirce wrote:

> ... implies that there are no things which belong to both classes at once.

However later in the lecture, he does in fact consider + between identical classes such as *a* + *a*, or *a* AND *a* as taking all the individuals of class *a* and then counting them all again.

Boole addressed the problem of addition of qualities by stating on page 30 of *Laws of Thought*:

> When I say, let *x* represent GOOD, it will be understood that *x* only represents GOOD when a subject for that quality is supplied by another symbol, and that, used alone its interpretation will be GOOD THINGS.

Echoes of this definition can be found in "Harvard Lecture III" on page 190 when Peirce assigned to *b* the meaning ALL BLACK THINGS.

For Boole, algebraic equality is now used to mean equivalence or identity. Equality in the sense of equivalence is stressed as a class concept in *Mathematical Analysis of Logic*. On page 24 we have:

> The general equation $x = y$ implies that the classes X and Y are equivalent, member for member; that every individual belonging to the one belongs to the other also.

This contrasts with *Laws of Thought* where equality is defined briefly as the on page 27 and later on page 34 the equality symbol, $=$, is regarded as a relation with which propositions are formed. In "Harvard Lecture III," Charles Peirce defined equality as, not merely identity with respect to number but, as he phrases it, complete identity without emphasizing the class concept. For example,

$$w = u$$

WASHINGTON CITY IS THE CAPITAL OF THE UNITED STATES

Here the is is the is of predication. This clearly follows *Laws of Thought* more closely than *Mathematical Analysis of Logic*.

While subtraction is not defined formally in *Mathematical Analysis of Logic* it is taken as exclusion in *Laws of Thought*.

On page 34 of *Laws of Thought* Boole wrote:

> Thus if x be taken to represent men, and y Asiatics ..., then the conception of ALL MEN EXCEPT ASIATICS will be expressed by xy. ...

He drew on analogy with transposition in algebra to state that if $x = y + z$ then $x - z = y$. In other words that subtraction is the inverse operation to addition. Subtraction is explained by Peirce in his "Harvard Lecture III" more clearly as

$$c|_a = b,$$

where b is the class c after the class a is taken away; in today's parlance, class complement, $c|_a$. Peirce also stressed the fact that this definition of subtraction assumes that c contains both a and b which is not explicit in Boole.

For the operation of negation we have in on page 20 of *Mathematical Analysis of Logic* the Duality Law for addition introduced in words only, viz.:

The class X and the class not-X together make the Universe.

A similar definition is given in *Laws of Thought*. However he did not express this equation in a mathematical form as Peirce did.

Regarding class complementation, Peirce first introduced $\bar{b}$ to denote everything NOT BLACK on page 193 of "Harvard Lecture III" and explicitly states the Duality Law for addition and from this derives the more usual notation for $\bar{b}$, i.e., $1 - b$.

Since $\bar{b} + b = 1$ we get $\bar{b} = 1 - b$.

Using the notation of $1 - b$ Peirce obtained the law of duality: $b(1 - b) = 0$. He claimed that this equation implies that b has only two numerical values, one and zero. This fact was also achieved, as shown earlier, by an argument which involved the use of Boole's addition operation with identical classes and the inference of $a = 0$ from the assumption $2a$ is well-formed, rather than from the equation $2a = 0$. The notation $\bar{b}$ for $1 - b$ is completely absent from *Mathematical Analysis of Logic* but first appeared in an example in *Laws of Thought* on page 119 in the interest of simplicity.

Let us turn our attention to the empty class and the universal class. In *Mathematical Analysis of Logic* there is no explicit definition of zero, but in *Laws of Thought* on page 47 Boole corrects this as follows:

> ... we must assign to the symbol 0 such an interpretation that the class represented by $0y$ may be identical with the class represented by 0, whatever the class y may be. A little consideration will show that this condition is satisfied if the symbol 0 represent nothing.

Charles Peirce followed the lead given in *Laws of Thought*. He defined 0 on page 190 in "Harvard Lecture III" as an entity which obeys the following rules:

$$h + 0 = h$$
$$c + 0 = c$$

Or ALL HORSES together with naught constitute ALL HORSES. And ALL cows together with naught constitute ALL COWS.

He continued:

> ...plainly means nothing; not nothing in respect to one measure merely as it does in arithmetic but absolutely nothing.

It is interesting to note that Boole goes further than Peirce by identifying 0 with the empty class. On page 47 of *Laws of Thought* he stated, "In accordance with a previous definition, we may term Nothing a class." But Peirce does not explicitly state that 0 is a class.

Regarding unity, in *Mathematical Analysis of Logic* the number 1 is defined straightforwardly as:

> ...the Universe, and let us understand it as comprehending every conceivable class of objects whether actually existing or not.

Later for secondary propositions, propositions that are true or false, Boole defined 1 as representing "all conceivable cases and conjunctures of circumstances," on page 49 of *Mathematical Analysis of Logic*. It is noticeable that Boole does not specify explicitly that 1 is a class.

The definition given in *Laws of Thought* on page 48, is similar except that here 1 is definitely a class:

> A little consideration will here show that the class represented by 1 must be "the Universe," since this is the only class in which are found all the individuals that exist in any class.

Here Boole used 1 as a unit of multiplication obeying the following rule:

The symbol 1 satisfies in the system of Number the following law, viz., $1 \times y = y$ or $1y = y$

and then used the above reasoning to show that 1 must be the Universe rather than defining 1 as the Universe first as in *Mathematical Analysis of Logic*. However in *Laws of Thought*, Boole followed De Morgan's concept of the Universe of discourse where the Universe is restricted to those existing objects under discussion. This differs from 1 in *Laws of Thought* which encompasses all objects whether existing or not.

Charles Peirce followed the method of *Laws of Thought* in looking at a set of equations or rules for 1 and then obtaining the definition of 1 from it. Drawing on analogies with arithmetic and algebra, he gave these examples in "Harvard Lecture III":

$$h \times 1 = h$$
$$c \times 1 = c$$

Now one ... is that class which has the whole of the objects of every class under it. In other words it is everything or whatever is.

Peirce then defined 1 or One as ALL THAT IS, and later as ALL THINGS. In contrast to his earlier definition of 0, 1 is definitely a class, as can be seen when he wrote "... one represents that class which multiplied by any other gives that other." 1 is also used in *Mathematical Analysis of Logic* in propositional logic as expressing truth. On page 51, we have "The elective symbol x selects all those cases in which the proposition is true, and therefore if the proposition is false $x = 0$."

Later on page 166 of *Laws of Thought* Boole was to write when discussing the Duality Law:

> ... in the expression of secondary propositions, 0 represents nothing in reference to the element of time. ... in the same system 1 represents the Universe, or whole of time, to which the discourse is Supposed in any manner to relate even 'eternity' unless some limitation is expressed or implied in the nature of the discourse.

With regard to these philosophical definitions of 1, all philosophical approaches to logic including the use of algebraic symbols to represent "the time for which propositions are true" or "the cases for which propositions are true," are absent in Peirce.

However such a use of 1 as in *Mathematical Analysis of Logic*, where the Universe is not defined as the Universe of Discourse but instead as simply EVERYTHING and in fact as used by Peirce in "Harvard Lecture III," leads to paradoxes and renders it impossible to distinguish between truth and tautological truth. It is noticeable that Peirce manages to avoid using the word UNIVERSE in his definition in "Harvard Lecture III," preferring instead the term EVERYTHING or WHATEVER IS. He also uses the terms ONE and the symbol 1. In following *Mathematical Analysis of Logic* here and not *Laws of Thought*, Peirce made a serious error. The Universe of Discourse is clearly not the class that Peirce as "the class which has the whole of the objects of every class under it."

3.3.3. Peirce on Numerical Values and Expansions.

Another point to consider is the interpretation of numerical values by Boole and how Peirce then used this concept in his own work. The topic of numerical values is not discussed in *Mathematical Analysis of Logic* but in *Laws of Thought*, Boole concluded through the Index Law $x^2 = x$ that there are only two possible numerical values in his algebraic logic namely 0 and 1. Peirce used a different approach. In "Harvard Lecture III," he drew on Boole's definition of addition of disjoint classes to show that all numbers beside 1 and 0 "mean that that which they are multiplied by is nothing."[36] However he applied this operation to identical classes which is clearly distinct from the logic of Boole which explicitly ruled out the application of this operation to all but disjoint classes. Peirce uses an analogy with language:

> ... you cannot say HORSES and HORSES BESIDES; although you can say NOTHING and NOTHING BESIDES. And, therefore, if you meet with such an expression as $a + a$, or a and a besides, you may be sure that a is nothing ...
>
> As $a + a$ makes $a = 0$ so does $a + a + a$, $a + a + a + a$ and so forth or in other terms $2a$, $3a$, $4a$ make $a = 0$ &c. Now this determines the meaning of all the other numbers besides one and zero. These numbers mean that that which they are multiplied by is nothing.

---

[36]See [**55**, p. 192].

In other words, $2a$, $3a$, etc. imply that $a = 0$. So Peirce showed that in Boole's calculus the only possible numerical values are 1 and 0. Peirce's stratagem is to allow addition between identical classes $a + a$ only if $a = 0$. This is a daring development of Boole, who in fact clarified his position in his correspondence with Jevons well after the publication of *Laws of Thought*. Boole held that $x + x$ was an uninterpretable symbol in logic, only interpretable in equations such as $x + x = 0$, in which case by the Index law this implied $x = 0$.

Peirce's departure was to consider $x + x$ as an expression leading to the inference $x = 0$. Also for Boole, in *Mathematical Analysis of Logic* and *Laws of Thought*, expressions such as $a + a$ or $2a$ mean different representatives of the same class. In his manuscript notes N7–N27 held in the Royal Society, which probably date from 1848 and now published in [**75**, p. 44] Boole wrote:

> ...we have $x + x = 2x$, $x + x + x = 3x$ whence also the idea of number ...it must be supposed that the $x1$ in $x1 + x1 + x1 + \ldots$ refer to different or mutually exclusive entities so that we may have the possibility of aggregation.

Peirce has inferred $a = 0$ from the assumption $2a$ is well-formed rather than the equation $2a = 0$ as Boole would have done. This is in contrast to Boole's second theorem of interpretation which states that for equations of the form $\varphi(x, y, z) = w$, where w is a constant, we are permitted to equate separately to 0 every term in which the coefficient does not satisfy the Index Law. Peirce seems to apply this to expressions of the form $\varphi(x, y, z)$ rather than equations.

The conclusion that numerical values are restricted to 0 and 1 is reached in *Laws of Thought*, through the Index Law and Peirce does indeed also use this form of the argument when discussing subtraction as will be seen later.

Let us consider more closely the use of $0/0$ in *Mathematical Analysis of Logic* and *Laws of Thought*. $0/0$ is introduced in the following way in *Mathematical Analysis of Logic*, page 72: Suppose we have $f(xy) = 0$, from the Development Theorem given on page 61 of *Mathematical Analysis of Logic*, which was obtained by Boole from analogy with analysis using Maclaurin's theorem. We have

$$f(x) = f(1)x + f(0)(1 - x),$$

with the version for a function of two variables being

$$f(0,0)(1-x)(1-y) + f(0,1)(1-x)y + f(1,0)x)1-y) + f(1,1)xy = 0.$$

The general expression of y as a function of x is therefore

$$y = vx + v'(1-x)$$

where

$$v = \frac{f(1,0)}{f(1,0) - f(1,1)} \text{ and } v' = \frac{f(0,0)}{f(0,1) - f(0,0)}.$$

Here since $f(1,0)$, etc. are numerical constants we may have $v = 0/0$ or $1/0$. On page 74 of *Mathematical Analysis of Logic*, we have, "In the former case, the indefinite symbol $0/0$ must be replaced by an arbitrary elective symbol v."

However, when v was first introduced on page 22 it is clear that v was intended to represent 'some'. It is implicit that v is not zero. Charles Peirce's definition is closer in spirit to *Laws of Thought*, which stresses the interpretation of $0/0$ rather than how it is reached.

$0/0$ is introduced on page 74 of *Laws of Thought*, as the symbol of indeterminate quantity which has, it is said, "a very important logical interpretation." He continued on page 89:

> Now, as in Arithmetic, the symbol $0/0$ represents an indefinite num-
> ber, ... analogy would suggest that ... the same symbol should rep-
> resent an indefinite class. '

An interpretation is reached on page 90 of *Laws of Thought*, in the following way: Suppose we have

$$y(1-x) = 0,$$

where y stands for all men and x stands for all mortals. Then,

$$y - yx = 0, \quad yx = y, \quad \text{and} \quad x = y/y.$$

But division, apart from division by 1 or 0, is not defined by Boole – especially not as the inverse of multiplication – although it is implicit in some of his examples; so to consider y/y we may use the development theorem and then look for an interpretation:

$$x = y/y = 1/1y + 0/0(1-y).$$

From this it is clear that 0/0 indicates that all, some or none of the class to whose expression it is affixed must be taken. On page 90 of *Laws of Thought*:

> We may properly term 0/0 an indefinite class symbol, and may if convenience should require, replace it by an uncompounded symbol v, subject to the fundamental law, $v(1 - v) = 0$.

Charles Peirce does not however, mention the symbol v in "Harvard Lecture III," but leaves this for "Harvard Lecture VI," where we shall be considering it further. Following on from this, Peirce arrived at the Boolean coefficient of 0/0 given in "Harvard Lecture III":

> h which stands for the class of all horses, as it is neither 0 nor 1, has no numerical value, i.e., it has a value which is not numerical.

Peirce calls this a "very peculiar and interesting point." However h can be represented by 0/0:

$$h \times 0 = 0.$$

Dividing both sides by zero we have,

$$h = 0/0.$$

Since this is true for any class we have 0/0 representing an indeterminate class.

Peirce justified this by analogy with arithmetic where division by zero is indeterminate. It is interesting to note that this early introduction of division as the inverse process of multiplication is another proof that he is already feeling his own way with Boole's algebraic logic and developing it along lines that are completely opposed to Boole's own theory. The operation of division as introduced in "Harvard Lecture VI" and Boole's own view of division is discussed more fully later.

In "Harvard Lecture III," Charles Peirce now moved away from *Laws of Thought* and *Mathematical Analysis of Logic*, to introduce completely new symbols 1/1 for 1 and 0/0 for 0. These new symbols neatly tie in with the existing 0/1 and 1/0. Peirce probably introduced them for symmetrical considerations, as with these symbols every expression $\varphi(m)$ can then be denoted by $Am + B(1 - m)$, where A and B are one of 1/1, 0/1, 0/0, 1/0.

Peirce gave a concrete example of the interpretation of an expression, e.g.,

$$\varphi(m) = m/m.$$

From Boole's Development Theorem we have,

$$\varphi(m) = \varphi(1)m + \varphi(0)(1 - m),$$
$$\text{so} \quad \varphi(m) = 1/1m + 0/0(1 - m).$$

It is then possible to arrive at an interpretation:

m/m stands for ALL MEN AND SOME, ALL OR NONE OF THE THINGS NOT MEN.

As mentioned earlier, division as a logical operation in *Laws of Thought* and *Mathematical Analysis of Logic* was strictly excluded. Boole was only interested in arriving at a logical interpretation for terms such as y/x for each particular problem. However as shown in [**133**, pp. 562–569], division is implicit in both *Mathematical Analysis* and *Laws of Thought* and even as in the following example taken from page 34 of *Mathematical Analysis of Logic*, more than implicit:

> A convenient mode of effecting the elimination, is to write the equation of the premises, so that y shall appear only as a factor of one member in the first equation, and only as a factor of the opposite member in the second equation, and then to multiply the equations, omitting the y. This method we shall adopt.

Peirce also followed Boole in not defining division as a logical operation in "Harvard Lecture III," although this was not the case as we shall see in "Harvard Lecture VI," and used Boole's method in arriving at an interpretation for m/m, using the Development Theorem as shown in the previous concrete example. However he then goes on to use division as an operation quite explicitly in checking his previous result. His argument given at [**55**, p. 196] is as follows:

$$am = m$$

That is, ALL MEN WHO ARE ANIMALS ARE THE SAME AS ALL MEN. Now divide by m and we have

$$a = m/m$$
but ALL ANIMALS ARE $a = m + 0/0(1 - m)$
and therefore $m/m = m + 0/0(1 - m).$

Peirce realized that the lack of a well-defined operation of division was a serious omission in Boole and although it was one of the three main differences between algebra and logic for Boole, Peirce now felt the need to supply such an operation and did so explicitly as we shall see later in "Harvard Lecture VI." Boole was mainly interested in the interpretation of x/y as a class and was reluctant to consider it as the inverse operation of multiplication in *Mathematical Analysis of Logic* and *Laws of Thought*.

3.3.4. Peirce on Elimination. Another feature of "Harvard Lecture III" is Boolean problem-solving, in particular, elimination of variables. This topic is not treated explicitly in *Mathematical Analysis of Logic* but in *Laws of Thought*, page 101, we have for the first time a method of elimination:

> ...the complete result of the elimination of any class symbols, xy, etc., from any equation of the form $V = 0$, will be obtained by completely expanding the first member of that equation in constituents of the given symbols, and multiplying together all the coefficients of those constituents, and equating the product to 0.

This is proved in Proposition I:

> If $f(x) = 0$ be any logical equation involving the class symbols x, with or without other class symbols, then will the equation $f(1)f(0) = 0$ be true, independently of the interpretation of x; and it will be the complete result of the elimination of x from the above equation.

Boole's main proof is as follows. Developing $f(x) = 0$ we have,

$$f(1)x + f(0)(1 - x) = 0$$

From this we obtain

$$x = \frac{f(0)}{f(0) - f(1)} \text{ and } 1 - x = \frac{f(1)}{f(0) - f(1)}.$$

Substituting into the fundamental law of logical symbols, $x(1 - x) = 0$, we get

$$f(1)f(0) = 0.$$

Peirce very briefly considers elimination on page 198 of "Harvard Lecture III" or as he phrased it: "we wish to be able to strike any letter out of an equation."

He used the method of elimination given in *Laws of Thought*. First he expands the left hand side of the equation completely using the Development Theorem. Then he multiplied the coefficients together and equated the result to zero. However at no stage does he give any explanation of the method he is using, or the procedure that he is following. The example given is $ab + c(1 - b) = 0$, to eliminate b.

Let us look more closely at Peirce's problem-solving methods, firstly concentrating on how he uses the Boolean Development Theorem. Start by letting

$$f(a, b, c) = ab + c(1 - b) = 0.$$

Then using the Development Theorem to expand the left hand side with respect to a and c:

$$f(1, b, 1)ac + f(1, b, 0)a(1 - c) + f(0, b, 1)(1 - a)c + f(0, b, 0)(1 - a)(1 - c) = 0$$

$$(b + (1 - b))ac + ba(1 - c) + (1 - b)(1 - a)c + 0(1 - a)(1 - c) = 0$$

(1) $$1ac + ba(1 - c) + (1 - b)(1 - a)c + 0 = 0$$

The required coefficients are now obtained by first substituting $b = 0$ and then $b = 1$ in the left hand side of this equation. We obtain, putting $b = 0$,

$$ac + (1 - a)c = c$$

Then putting $b = 1$, we obtain from equation (1):

$$ac + a(1 - c) = a.$$

Multiplying these two coefficients together and equating the product to zero,

$$ac = 0$$

thus eliminating b.

Peirce then produces the following alternative version on page 198 of "Harvard Lecture III." This uses the method of elimination given by Proposition 1, where it is unnecessary to expand the function in terms of a and c. As elimination appears in *Laws of Thought*, only and not *Mathematical Analysis of Logic*, this is another instance of how Peirce follows the 1854 version of George Boole's algebraic logic. Charles Peirce's alternative version is as follows:

> Now this result may be got also by writing
> $$ab + c(1 - b)$$
> Put $b = 1$ and $b = 0$
> $$ac = 0.$$

This rather cryptic explanation can now be seen to follow Boole's method of elimination as given in Proposition 1, *Laws of Thought* on page 101 when written in the following way. Let
$$f(a, b, c) = ab + c(1 - b) = 0;$$
Then
$$f(a, 1, c).f(a, 0, c) = 0$$
so
$$ac = 0.$$
At the end of this section of "Harvard Lecture III," on page 199, Charles Peirce finally attempts to use Boole's algebraic logic on a classical argument expressed in syllogistic form but in fact he does not complete the example and fails to draw the necessary inference. Peirce only provides an expression of the argument in the form of an algebraic equation which he unfortunately does not make any attempt to solve. His expression of the problem is as follows:

> Now let us put an argument into syllogisms.
>
> All men are Animals $\dfrac{m(1 - a)}{ma + m(1 - a)} = 0$
>
> Socrates is a Man $\dfrac{s(1 - m)}{sm + s(1 - m)} = 0$
>
> $$\dfrac{m(1 - a)}{ma + m(1 - a)} + \dfrac{s(1 - m)}{sm + s(1 - m)} = 0$$
>
> As I know your minds must be wearied with this mathematics, I will now postpone the further consideration of it for another lecture and will take up now a lighter subject.

The complexity of this method of resolving syllogistic logic ensured that Peirce did not proceed further in this treatment of classical syllogisms. Previously on page 198 Peirce states:

> ... if we have such an expression as the following
> $$\frac{xy + x(1 - y) + (1 - x)y + (1 - x)(1 - y)}{xy + x(1 - y) + (1 - x)y + (1 - x)(1 - y)} = 0$$
> or any other expression derived from this by striking out any term or terms from numerator or denominator or both ...
> 1) No individual among all the classes appearing in the numerator exists
> 2) Some individual among all the classes not in the numerator but in the denominator exists.

Following this form, Peirce's expression of ALL MEN ARE ANIMALS as
$$\frac{m(1 - a)}{ma + m(1 - a)} = 0$$
instead of the simpler $m(1 - a) = 0$ has the advantage that this implies both that there are no men who are not animals but also that the class of men who are animals exists. Simplifying the equation
$$\frac{m(1 - a)}{ma + m(1 - a)} + \frac{s(1 - m)}{sm + s(1 - m)} = 0$$
we have,
$$\frac{m(1 - a)(sm + s(1 - m)) + s(1 - m)(ma + m(1 - a))}{(ma + m(1 - a))(sm + s(1 - m))} = 0.$$
and expanding the numerator, we have
$$\frac{m(1 - a)sm + m(1 - a)s(1 - m) + s(1 - m)ma + s(1 - m)m(1 - a)}{(ma + m(1 - a))(sm + s(1 - m))} = 0.$$
Using the Index Law $x^2 = x$ and the Duality Law $x(1 - x) = 0$ this simplifies to
$$\frac{m(1 - a)}{mas} = 0.$$
This implies that there is no individual in the class consisting of common members of the class containing SOCRATES, the class containing ALL MEN and the class of NON-ANIMALS and furthermore the class consisting of common members of the class of MEN, SOCRATES and ANIMALS exists. In other words SOCRATES is an ANIMAL.

In conclusion, "Harvard Lecture III" is an attempt by Charles Peirce to introduce George Boole's algebraic logic as developed in *Laws of Thought* rather than *Mathematical Analysis of Logic* in a simple way, omitting problem areas such as the interpretation of v, any philosophical notions of cases of propositions or periods of time and any discussion of secondary propositions. The concept of the Universe of Discourse is also omitted as are any examples in the application of algebraic logic to problems of probability. However there are novel ideas put forward by Peirce such as the notation of 0/1 for 0 and 1/1 for 1 and the expression of propositions in the form of an equation:

$$\frac{xy + x(1-y) + (1-x)y(1-x)(1-x)}{xy + x(1-y) + (1-x)y + (1-x)(1-y)} = 0.$$

This emphasized the existence of classes, a concern Peirce was to return to in "Harvard Lecture VI."

The explicit and early use of the logical operation of division by Peirce, which is absent from Boole's *Mathematical Analysis of Logic* and *Laws of Thought*, occurs in "Harvard Lecture III," although it is not defined and again this is something Peirce is to develop in "Harvard Lecture VI." The treatment given by Peirce in this introductory lecture is entirely uncritical. He believed that *Laws of Thought* was destined "to mark a great epoch in logic." This unquestioning acceptance is to change when we consider "Harvard Lecture VI," which emphasizes the defects of Boole's calculus of logic and makes some attempts at improving and supplying any deficiencies of the algebraic logic. However this trend can be perceived as early as "Harvard Lecture III," which purports to introduce Boolean algebraic logic, and yet either Peirce is misinterpreting key concepts of the Boolean calculus when he introduces such concepts as:

1) the use of $a + a$ to represent addition between identical classes in the expression $2a$, as opposed to Boole's view of $2a$ as aggregation between distinct representatives or members of the same class and only logically valid when contained in equations of the form $2a = 0$,

2) the inference of $a = 0$ from the expression $2a$ rather than, as in Boole's second theorem of interpretation, from the equation $2a = 0$ and

3) the use of division as a logical operation; or it may be the case that he cannot even at this early stage resist the temptation of developing and extending Boole's algebraic logic.

3.3.5. Peirce's "Harvard Lecture VI" (1865). In the introduction to "Harvard Lecture VI," Charles Peirce described Boole's work on algebraic logic as "the most extraordinary view of logic which has ever been developed with success." He then drew attention to the two sorts of symbols in use at the time to represent logical processes. These were geometrical and algebraic symbols.

FIGURE 3.2.   Geometrical Euler Circles: Subordination

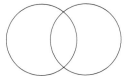

FIGURE 3.3.   Geometrical Euler Circles: Logical Intersection

Algebraic as used by Sir William Hamilton and Gottfried Ploucquet:[37]

FIGURE 3.4.   Every man is a creature.

Peirce recognized these algebraic notations as superior to the geometrical ones because they involve some analysis of the laws of logic. However he argued that both systems are "utterly useless . . . in practice and as the basis of a conception of the science." He also claimed that Boole's calculus combined the best features of the two

---

[37]Although one of the 18[th]th century's major contributors to the development of formal logic, Ploucquet's work seems to have been little known in general. C. I. Lewis could write in [**115**, p. 18] that copies of Ploucquet's books were unavailable in U. S. A. and that "attempts to secure them from the continent have so far failed."

FIGURE 3.5.   Some man is not an Ethopian.

FIGURE 3.6.   Some creature is not an Ethopian.

notations resulting in a new relationship between logic and mathematics. In "Harvard Lecture VI" we have:

> For like the literal notations it is abstract and deals with the laws of logic themselves and like the geometrical notations it brings out a harmony between logic and mathematics, so as to render the former easier to think about ... it reflects upon mathematics a new light from logic.[38]

On the other hand he was very far from saying that Boole's algebraic logic is a perfect representation of logic. This is a radical shift from the uncritical stance of his introductory "Harvard Lecture III." He now sought to add to the notation and fill the "gulfs ... which were entirely overlooked by its author." Furthermore he argued that logical judgements are not satisfactorily expressed by algebraic logic. He continued by saying "but a very small fraction of all judgements can be expressed in this way." For Peirce the attempt by Boole to express all the laws of reasoning in terms of algebraic notation has ended in failure. He wrote of "enormous deficiencies" but considered that "the method is in its infancy" and obviously hoped to continue its development.

A brief description of Boole's calculus of logic follows with definitions of equality, addition and subtraction similar to the definitions in "Harvard Lecture III." It is interesting to note that Peirce made it explicit that for subtraction $x - y$ clearly means

---

[38]See [**143**, p. 225].

that x contains y "for otherwise $x - y$ would be an absurdity and incapable of interpretation." Peirce now defined the rule of transposition,

$$a + b = c \ \text{ then } \ c - a = b$$

where a and b are disjoint classes and used this not only to define subtraction but also zero as the class that obeys $a - a = 0$.

In Boole's algebraic logic, the missing operation is division. For the first time in his Harvard Lectures, Peirce now defined the logical operation of division, something that Boole excluded from his calculus. He defined it in the following way:

$$\text{From } ax = b \text{ we have } x = \frac{b}{a}.$$

Peirce called this the rule for clearing from fractions and so obtains from this rule his definition of division. He defined b/a as "a class which includes b and nothing but b that is at the same time a; that is, it comprises all b and some, all or none of what is not a besides." It is clear that a contains b. This is represented by the diagram Figure 3.7 below. This cannot be represented in current notation by

$$x = \frac{b}{a} = b + \bar{a}.$$

because of the indeterminate class SOME, ALL OR NONE of a. Instead it is possible to represent it by

$$x = \frac{b}{a} = b + \frac{0}{0}\bar{a}$$

or

$$x = \frac{b}{a} = b + \frac{0}{0}(1 - a)$$

Compare this with [**115**, p. 81] which gives a similar definition in the following way: b/a has lower limit b and upper limit $b + a$, and $b/a = ba + v(ab) + [0](b - a)$, where [0] indicates that the term to which it is prefixed must be null. This gives the same result as the above definition for b contained in a. However Lewis has incorrectly given that the corresponding condition should be that a is contained in b, whereas it should be that b is contained in a.[39]

It seems to be the case that Charles Peirce in trying to strive for analogy with arithmetic rather labors the connection. Although for subtraction, the analogy with arithmetic holds, i.e. for $x - y$, the class x contains the class y, it is not true that for

---

[39]Lewis uses a definition for a/b, which has been transposed for clarity.

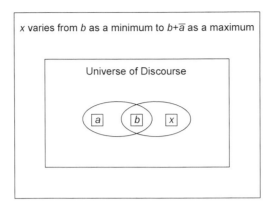

FIGURE 3.7.   Universe of Discourse

b/a then b contains a, rather as we have shown in Figure 3.7, a contains b. Peirce does not argue that b does contain a in his discussion on page 227 of "Harvard Lecture VI," rather he states that the meaning of b includes the meaning of a. But this is true since the class a contains the class b.

Peirce uses the terms *extension* to refer to the class of objects and *comprehension* or *intension* for class description:

> At the same time just as the inverse process of subtraction implies in itself that the extension of the subtrahend includes the extension of the minuend; so the inverse process of subtraction implies in itself that the extension of the subtrahend includes the extension of the minuend; so the inverse process of division implies that the comprehension of the dividend includes the comprehension of the divisor. Thus take x/y. Now unless x/y contains in itself y as a factor the division cannot be performed and the expression is incapable of interpretation.

Peirce is claiming here that the meaning of x must encompass the meaning of y. But this is the case since as we have seen y contains x. For example let y be the class of all animals and let x be the class of cats. Then x/y is the class of cats and some, all or none of the class of non-animals, and the meaning of cat does include the meaning of animal. This rule for clearing from fractions is also used to define 1, as $b/b = 1$. It seems that he is now introducing a number of rules, i.e., the rule of transposition and the rule for clearing from fractions and so obtaining his definitions from these rules, rather than as was the case in "Harvard Lecture III," obtaining his definitions from the logical operations themselves.

3.3.6. Peirce's Logical Laws. An important section of "Harvard Lecture VI" discusses the three fundamental laws of logic, the Law of Identity, the Law of Excluded Third and the Law of Contradiction on page 227. Peirce's aim is to show that the first two laws are not satisfactorily expressible in Boole's algebraic logic. Let us firstly consider identity. The Law of Identity is ALL A IS A. Peirce shows that this is expressed in Boole's calculus as $A = A^2$. However he does not consider that this implies existence, i.e., A *is* something, or in other words that every logical term is capable of an affirmative predicate or that it is *real*, since A *is* A does not imply the existence of A but means nothing more than A is X. He instead proposes that the closest approximation to this law expressible in Boole's calculus states that all the individuals composing a class have the class-character, which he shows in the following way:

(1) Let A denote a certain class.
(2) Let a denote the individuals in A.
(3) Let α whatever has the class-character.
(4) Then we have $A = a\alpha$.
(5) But since these are all identical we have $a = a^2$.

It is interesting to see that Peirce arrives at the definition of an idempotent basis that as we have seen was one of the fundamental concepts of his father Benjamin Peirce's *Linear Associative Algebra*. However this predates the lithographic version by five years.

We shall now consider the Law of Excluded Third which states that A is either B or *not-B*. This cannot be expressed in its normal meaning of A either is B or is *not-B* because of the problem of showing the existence of A, but if the law means A either is$B$ or is *not-B* then this can be expressed in the following equation:

$$(a - 1)(a - 0) = 0.$$

The Law of Contradiction is also expressible in Boole's calculus, as the law means A is *not-not-A* or what comes to the same thing A which is *not-A* is non-existent, which is expressible as $a(1 - a) = 0$. Peirce concluded by noting that the three fundamental laws are algebraically identical.

He next considers some other defects of Boole's calculus in relation to the categories of judgements that can be expressed. In this he shows the influence of Kant and Hegel in his discussion. I have summarised the judgements in Figure 3.8.

Peirce noted that Apodeictic Judgements, IF X THEN Y cannot be distinguished from Assertory Judgements, X IS Y, in Boole's system. As a result neither Hypothetical Judgements, IF A IS B THEN C IS D, or Disjunctive Judgements, EITHER X IS TRUE OR Y IS TRUE can be expressed with ease.

The examples that Peirce gives are as follows:

a Problematic Judgements, by which Peirce means propositions expressing the possible or impossible, are not clearly expressed in algebraic logic unless xy is taken to mean x *may be* y.

b Hypothetical Judgements cannot be expressed except the following construct: Let a express THERE IS AN EAST WIND and let b express THE BAROMETER WILL RISE. Then $a = ab$ will mean IF THERE IS AN EAST WIND THE BAROMETER WILL RISE. Peirce points out that this alters the meaning of the sign of equality.

c Judgements of Quantity. The following example of Boole's is cited as fallacious:

$$vx = v(1 - y)$$

This is taken by Boole to mean SOME X IS NOT Y. Transposing we get

$$vy = v(1 - x)$$

which is SOME Y IS NOT X.

Peirce continued on page 231 of "Harvard Lecture VI":

But it does not follow from SOME X IS NOT Y that SOME Y IS NOT X. This expression is therefore wrong.

The difficulty as analyzed by Peirce is contained in the use of v to represent an indefinite class whereas it should be a particular class. He gave as an example:

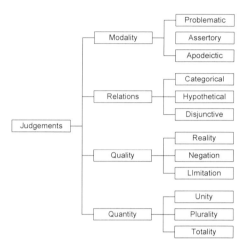

FIGURE 3.8.   Judgements

Let X be the class of all animals and Y be the class of all men.

Then it is clear that instead of v standing for an indefinite class it should be instead a class of animals. He reasoned:

> When we say "Some animals are not men," *some* is not a wholly indefinite class for it is understood to be a class of animals.

We now consider the use of the indefinite class v as used by George Boole in both *Mathematical Analysis of Logic* and *Laws of Thought*.

3.3.7. Boolean Quantification. Boolean quantification was mainly expressed by the arbitrary elective symbol v as defined in *Mathematical Analysis of Logic*, which owes its importance to the fact that it represents the key operation of inclusion in Boole's part/whole logic of classes; $y = vx$ means both *Some Xs are Ys* and and *All Ys are Xs*.

In modern notation for classes we have $Y \subseteq X$. Here it is implicit that v is non-empty. As is the case on page 22 of *Mathematical Analysis of Logic* when we have the first definition of v in MAL as $v \equiv xy$ Boole says:

> v includes all terms common to the classes X and Y, we can indifferently interpret it as SOME Xs or SOME Ys.

This is confirmed later when on page 27, under the heading "Of the conversion of propositions," where Boole draws a distinction between two primary canonical forms already determined for the expression of propositions:

No Xs are Ys	$xy = 0$	... E
Some Xs are Ys	$v = xy$	... I

This clear distinction between the definitions shows that in *Mathematical Analysis of Logic* Boole intended v to be an arbitrary elective symbol meaning SOME in the sense of AT LEAST SOME AND POSSIBLY ALL but not NONE.

Another instance where he distinguishes between v meaning SOME but not NONE occurs earlier on pages 23 and 24 of *Mathematical Analysis of Logic* which discusses the following example:

> To express the Proposition SOME Xs ARE NOT Ys write

(3.1) $$v = x(y - 1)$$

> Multiplying both sides by x we get

$$vx = x^2(1 - y) = x(1 - y) = v$$

> Alternatively multiplying both sides by $1 - y$

$$v(1 - y) = x(1 - y) = v$$

It is clear here that since $v(1 - y) = v$ we have $vy = 0$. Boole observes:

> Since in this case $vy = 0$, we must of course be careful not to interpret vy as *Some Y*s.

Thus implying that v does not encompass the meaning of NONE in its definition. On the other hand, v clearly does satisfy the Index Law. On page 72 of *Mathematical Analysis of Logic* Boole says:

> v should be an elective symbol ... (since it must satisfy the index law), its interpretation in other respects being arbitrary.

On page 63 of *Laws of Thought* v is defined as a class:

> ... indefinite in all respects but this, that it contains some members of the class to whose expression it is prefixed.

Later Boole replaces the indefinite symbol $0/0$ by v. This is permissible but in fact $0/0$ means more than v. For the first time v can take the value 0. $0/0$ is introduced in an example on page 76 of *Mathematical Analysis of Logic*:

> Given $x(1 - z) + z = y$, to find z.

Using the Development Theorem, Boole eventually arrived at the following equation:
$$z = 0/0xy + 1/0x(1 - y) + (1 - x)y$$
He then continued:

> ... the indeterminate coefficient of the first term being replaced by v, an arbitrary elective symbol, we have
> $$z = (1 - x)y + vxy$$
> the interpretation of which is, that the class Z consists of all the Ys which are not Xs, and an indefinite remainder of Ys which are Xs. Of course this indefinite remainder may vanish.

It is clear that v is "indefinite in the highest sense, i.e., it may vary from 0 up to the entire class." So when v is replacing $0/0$ it can take the value 0. v is first introduced on page 61 of *Laws of Thought* in the following way:

> Represent then by v, a class indefinite in every respect but this, viz.,
> that some of its members are mortal beings, and let x stand for MOR-
> TAL BEINGS, then will vx represent SOME MORTAL BEINGS.

This shows that *Laws of Thought* as did *Mathematical Analysis of Logic* assumes that v is non-empty. Again we have on page 63 of *Laws of Thought* emphasizing the fact that v does contain individual members:

> ... introducing v as the symbol of a class indefinite in all respects but
> this, that it contains some individuals of the class to whose expres-
> sion it is prefixed.

Later on page 87 of *Laws of Thought* 0/0 is established as an indefinite remainder clearly meaning SOME, NONE, OR ALL. This is then replaced by v the indefinite class symbol, whenever it occurs. However these two symbols are seen as distinct. Although v is subject to the Duality Law, we have on page 91 of *Laws of Thought*:

> The symbol 0/0, whose interpretation was previously discussed,
> does not necessarily disobey the law for it admits of the numerical
> values 0 and 1 indifferently.

Boole here is clearly cautious about committing himself as to whether 0/0 obeys the Duality Law. Further evidence to show this distinction between 0/0 and v occurs on page 124 of *Laws of Thought*:

$$vX = vY$$

> but v is not quite arbitrary, and therefore must not be eliminated. For
> v is the representative of SOME, which, though it may include in its
> meaning ALL, does not include NONE.

It is explicit that v does not mean NONE. However later when considering secondary propositions and the philosophical aspects of his work in *Laws of Thought* appears to have been a shift of position:

v is thus regarded as a symbol of time indefinite, *vx* may be understood to represent the whole or an indefinite part, or no part, of the whole time.

In summary, *v* is first introduced as meaning SOME, POSSIBLY ALL as a way of expressing the Aristotelian canon SOME *X*s ARE *Y*s. Later Boole introduced 0/0 as an indeterminate coefficient. Its interpretation is established as SOME, ALL OR NONE. This similarity with the indeterminate logical symbol *v* makes it possible for it to be replaced with *v* and so that logical expressions are presented solely in the form of logical elective symbols. However as we have seen, 0/0 represents more than *v* and so Boole expands his meaning of *v* to encompass the wider interpretation of 0/0 to include the meaning NONE. This distinction between the two symbols explains why Boole is unsure whether 0/0 obeys the Duality Law, whereas none being a logical symbol must do so. The inconsistencies in the use of none arise when none is taken to mean SOME, ALL OR NONE in the second half of both *Mathematical Analysis of Logic* and *Laws of Thought* when considering secondary propositions. This differs from the original use of *v* to represent SOME OR ALL BUT NOT NONE in primary propositional logic. Later in *Laws of Thought* he does not trouble to replace 0/0 by *v*, perhaps recognizing the narrower scope of *v*.[40] After having listed all the defects of Boole's algebraic logic, Peirce then points out its advantages including the fact that the difference between analytic and synthetic judgements is easily shown as is the difference between Disjunctives, Either *X* is true or *Y* is true, and Divisives, Either *X* or *Y* is true.

3.3.8. Early Use of Boolean Operations. Peirce continued "Harvard Lecture VI" on page 231 by noting that addition and multiplication are inverse operations or as he phrases it:

> It is curious that extensive combination should be represented by addition; and comprehensive combination by multiplication when the extensive and intensive quantities stand in reciprocal relation ... multiplication can undo the work of addition and vice versa.

He gave the following example:

---

[40]De Morgan in his *Syllabus of a Proposed System of Logic* [**38**] is clear that SOME denotes NOT NONE.

Starting with the sum of two classes a and b written as $a(1 - b)$ and $b(1 - a)$ to show they are distinct, i.e.,

$$a(1 - b) + b(1 - a).$$

Multiplying by a we have

$$a^2(1 - b) + ab(1 - a).$$

By the Index and Duality Laws this gives

$$a(1 - b),$$

which is now a product of two terms.

Surprisingly when Peirce goes on to consider a similar example with the product of two terms on page 232 of "Harvard Lecture VI," he makes a fundamental error in his reading of Boole's logic. He uses an equation of the form $x + x = x$ which is an equation that Boole himself rules out completely. Peirce has missed the fundamental concern of Boole that aggregation of members of the same class in logic is equivalent to the plural of a name in ordinary language.[41] In fact $2x$ is uninterpretable and $2x = 0$ implies that $x = 0$. Here is Peirce's example:

Taking a product of two classes,

$$xy$$

adding x,

$$xy + x.$$

However here

$$x = xy + x(1 - y),$$

so we have

$$xy + xy + x(1 - y).$$

Peirce now continues:

We have then $2xy + x(1 - y)$ but the coefficient means nothing. It may be struck off. We have then $xy + x(1 - y)$ or x.

----

[41] See [**133**, Section 8.2].

The error occurs when Peirce replaces $2xy$ with xy which is totally at odds with Boole's own views of the nature of numerical coefficients. He regarded numerical coefficients as a notation for the aggregation of distinct members of the same class.

On page 232 of "Harvard Lecture VI," Charles Peirce then proceeds to outline the two theorems that he declares are the foundation of the application of Boole's calculus to ordinary reasoning. These are the Development Theorem and the theorem for elimination given in Proposition 1 of *Laws of Thought*. He gives the Development Theorem in the following format:

$$fx = xf1 + (1 - x)f0.$$

He then used this theorem for two variables a and b in the following form:

$$f(a, b) = abf(1, 1) + a(1 - b)f(1, 0) + (1 - a)bf(0, 1) + (1 - a)(1 - b)f(0, 0),$$

when considering an interpretation for the expression $(1 - (a - b)^2)/a$. He arrived at:

$$(1 - (a - b)^2)/a = ab + 0/0(1 - a)b + ((1 - a)(1 - b))/0$$

Here $0/0$ is interpreted as an indeterminate class meaning ALL, SOME OR NONE SO

$$0/0(1 - a)b$$

means SOME, ALL OR NONE OF b WHICH IS NOT a.

Peirce then attempted to interpret $((1 - a)(1 - b))/0$ using his definition of division in which the dividend contains the divisor as a factor to arrive at the conclusion that

$$(1 - a)(1 - b) = 0.$$

Not satisfied with this, he also gave an alternative argument using multiplication by zero. Both these arguments are given on page 233:

> To find what $((1 - a)(1 - b))/0$ means we must remember that when one letter is divided by another, as x/y, it must be that the dividend contains the divisor as a factor. Now to say that $(1-a)(1-b)$ contains zero as a factor is to say ... it enters into the meaning of nothing and does not exist. This same result may also be obtained thus.
>   Let $((1-a)(1-b))/0 = y$; then multiplying by zero $(1-a)(1-b) = 0$, while the value of y is wholly indeterminate.

The statement "$(1 - a)(1 - b)$ contains zero as a factor" refers to the condition that $(1 - a)(1 - b) = 0y$ for some y, not that $(1 - a)(1 - b)$ contains zero.

nt = 0	The ancestors of Negroes had no tails.
m=tm	Monkeys have tails
nt + m(1-t)=0	
mn = 0	None of the ancestors of Negroes were monkeys.

The above arguments would have been simplified by using the definition of Peircean division so that 0 contains the class $(1 - a)(1 - b)$ as shown earlier; so $(1 - a)(1 - b)$ must be 0.

Peirce finished the algebraic logic section of "Harvard Lecture VI" by considering how a number of different equations may be combined into a single equation. He gave a simple example on page 235 where two equations are given and transposed to equal zero. After squaring if necessary, the equations are added together. Elimination can now be effected using Proposition 1. At this stage we have seen that Peirce gives a very close interpretation of Boole's algebraic logic. The main differences falling into two categories.

1. *Philosophical.* Peirce treats only Boole's primary or categorical logic – propositions of the form X is Y – and thereby ignores Boole's secondary or hypothetical logic – propositions of the form IF A is B THEN C is D. In fact as we have seen he discards it, claiming that such hypotheticals cannot be expressed in Boole's logic.

One advantage of this is, of course, that Peirce does not have to discuss any of the philosophical interpretations of instances, cases, truth or falsehood, and periods of time that Boole attached to his logical symbols in his secondary propositional logic. However Peirce did consider the application of Boole's method to probability which contains such hypothetical propositions. In doing so he diverged from Boole in a marked way by allowing the variables s and t other values besides 0 and 1. He wrote on page 237:

> ... instead of insisting any longer upon allowing s and t only the two values zero and unity, we must allow them values proportionate to the number of cases in which they will occur. Then the probability of s will be represented by a fraction whose numerator is s and whose denominator is the sum of all the possible cases.

2. *Definitions and Rules of Logical Operations*. Peirce, especially in "Harvard Lecture VI," derived definitions from rules and laws more frequently than Boole who defined his operations and then obtained rules and laws subsequently. However the laws of commutativity and associativity are not included by Peirce in Harvard Lectures III and VI. On the other hand, in a revolutionary departure from Boole, he introduced and defined the operation of logical division, an operation clearly not countenanced by Boole. He also does not make use of the logical symbol v, instead preferring the less ambiguous 0/0, thus avoiding the inconsistencies discussed earlier, that arose when Boole used v in the sense of 'some or all' and 0/0 in the sense of SOME, ALL OR NONE.

The algebraic logic of Boole, in particular as expounded *Laws of Thought* is followed fairly faithfully in "Harvard Lecture III" apart from the two main areas of divergence. However by "Harvard Lecture VI," Peirce is pointing out defects and omissions from Boole's work and trying to supply and improve the calculus. This is manifested in the definition of the operation of division as provided by Peirce and omitted by Boole. It is clear that Peirce has found in Boolean algebraic logic an exposition of logic in an algebraic form that he has absorbed completely – the only serious error being his misinterpretation of $2xy$ as xy – and is now eager to continue its development. This is shown in his next work on algebraic logic two years later.

3.4. Development and Expansion of Boolean Algebraic Logic

In Peirce's next paper, "On An Improvement in Boole's Calculus of Logic," the main shift from Boole appears to be away from operations between disjoint classes towards those operations in particular addition which no longer need such a qualification. This was first introduced by Jevons in his book *Pure Logic*, and I will now briefly outline its main trends.

3.4.1. Jevons's *Pure Logic*. In this work, properly entitled *Pure Logic or the Logic of Quality Apart from Quantity with remarks on Boole's system and on the Relation of Logic and Mathematics*, Jevons showed that he shared with Boole, his concern with the abstract, generalizing power of algebra when applied to logic.[42] stated his aims:

---

[42]See [**103**, p. 3].

> It is the purpose of this work to show that Logic assumes a new
> degree of simplicity, precision, generality, and power when compar-
> ison in quality is treated apart from any reference to quantity.

Jevons acknowledged his debt to Boole, frequently referring to LT *Laws of Thought*
throughout his work. It is interesting to note that Jevons wrote on page 5:

> The forms of my system, may in fact, be reached by divesting his
> system of a mathematical dress...

which is exactly what Boole himself tried to do after *Laws of Thought*. However
although founding *Pure Logic* on Boole's algebraic logic, Jevons has completely different ideas of the fundamental relationship between logic and mathematics. He not
only misunderstands Boole's view that logic and mathematics are separate branches of
a universal calculus of reasoning, but also stated his own logistic position. On page 5
we have:

> ...it may be inferred, not that Logic is a part of Mathematics, as is
> almost implied in Professor Boole's writings, but that the Mathemat-
> ics are rather derivatives of Logic.

Jevons's own position was that he thought of mathematics or arithmetic as a calculus
of quantities as opposed to logic, which he regarded as a calculus of qualities, so that
mathematics is dependent on logic, and he later wrote on page 77:

> Logic and mathematics are certainly not independent. And the clue
> to their connection seems to consist in distinct logical terms forming
> the units of mathematics.

Furthermore he was careful not to stress any mathematical terms and notation *Pure
Logic*, page 8, states:

> Let it be borne in mind that the letters *A*, *B*, *C* etc. as well as the
> marks +, 0 and =, afterwards to be introduced are in no way mys-
> terious symbols ... There is consequently nothing more symbolic or
> mysterious in this system than in common language.

Logical Propositions	Mathematical Equations
Terms known admit ...	*Numbers known admit ...*
Combination	Multiplication
Separation unless dividend contains divisor	Division unless divisor is 0
Terms unknown admit ...	*Numbers unknown admit ...*
Combination	Multiplication
... but do not admit Separation	*... but do not admit* Division

It is interesting to note that Boole also discarded the mathematical and algebraic reasoning of *Laws of Thought* in favor of developing his calculus without mathematical symbols. In the main body of *Pure Logic*, Jevons established the laws of commutativity and associativity. He then stated the Law of Simplicity, $AA = A$, referring to this as Boole's Law of Duality although it is in fact the Index Law. AB is defined to be the sum of the meanings of A and B. This is Boole's $A.B$ and today's intersection, $A \cap B$.

Page 21, of *Pure Logic* notes the difficulties associated with a logical operation of division, relating this to the restriction of division by zero in arithmetic, but unlike Peirce does not proceed to define such an operation. The following correspondences between logical propositions and mathematical equations are then listed thus establishing a clear relation between logic and mathematics:

He goes on to claim:

> The above analogies did not escape the notice of Professor Boole and I am therefore at a loss to understand on what ground he asserts that there is a breach in the correspondence of the laws of logic and mathematics.[43]

However Jevons conveniently forgets that he himself had pointed out one such breach between logic and mathematics on page 20:

---

[43]For the differences between Boole's calculus of logic and mathematics – of which Boole was well aware – see earlier in this section pp. 61–62.

It will be obvious that a mathematical term or quantity of several factors is strictly analogous in its laws to a logical combined term, excluding the *Law of Simplicity*[44]

On page 25, when discussing plural terms, Jevons introduced the Law of Unity, $A+A = A$ in the process noting "It was not recognized by Professor Boole, when laying down the principles of his system." In fact this law is specifically ruled out by Boole as his operation of addition applied only to distinct classes. Here for the first time, Jevons extended the logical operation of addition to classes that are no longer distinct. He used *term* in the sense of class, viz., a collection of individuals, and defined the Universe as:

> The sphere of an argument, or the Universe of Thought, contains all the included subjects.[45]

However he did not assign it a symbol, e.g., 1. Furthermore he continued on page 59:

> There is thus no boundary to the universe of logic. No term can be proposed wide enough to cover its whole sphere.

This shows that like Peirce he did not restrict it to a Universe of Discourse. Several definitions then follow, including the term 0 meaning EXCLUDED FROM THOUGHT and a for NOT-A or the class complement of A.

As we have seen earlier, Peirce claimed that Boole could not properly express SOME A. Jevons also recognized this problem and introduced the notation $A = AB$ for Boole's $A = vB$ meaning A IS SOME B. The motivation for this logic was, like that of Boole and De Morgan, to develop a method to encompass and express traditional syllogistic logic. Before considering a specific example of Jevons's logic applied directly to a problem taken from *Laws of Thought* let us summarize Jevons's main laws or as he called them the Conditions of Logic.

---

[44]Author's emphasis.

[45]See [**103**, p. 44].

Law of Sameness	$\{A = B = C\} = \{A = C\}$
Law of Simplicity	$AA = A, BBB = B$, etc.
Law of Same Parts and Wholes	$AB = BA$
Law of Unity	$A + A = A, B + B + B = B$, etc.
Law of Contradiction	$Aa = 0, Bb = 0$, etc.
Law of Duality	$A = A(B + b) = AB + Ab$, etc.
Law of Commutation[46]	$B + C = C + B$
Law of Association[46]	$A(B + C) = AB + AC$

3.4.2. Example: The Definition of Wealth. Jevons[47] makes a direct comparison with Boole's system as applied to a problem expressed by the following definition of wealth:

> Wealth is what is transferable, limited in supply, and either productive of pleasure or preventive of pain.

Using his method of obtaining inferences from propositions, Jevons obtained the solution using this alternative method. Jevons's method, which he called the Method of Indirect Inference, consists of writing down all possible combinations of the terms A, B, C, and D and their contraries a, b, c, and d and then combining each of these terms separately with both sides of a premise. Dual terms, e.g., $B + b$, may be struck out, as well as those obtained by intrinsic elimination where we may substitute for any part of one member of a proposition the whole of the other. For example, in $A = BCD$ if we wish to eliminate D we can write $A = ABC$, i.e., write A instead of D. Finally those terms that form a contradiction with one side of the premise are also eliminated.

Let A=Wealth, B=Transferable, C=Limited in Supply, D=Productive of pleasure, and E=Preventive of pain. Then the definition in question is expressed by the proposition

$$A = BC(DE + De + dE)$$

which includes all the combinations of D, E, d, e, except de because by definition wealth must be either productive of pleasure or preventive of pain. Since $E + e = 1$ and $A = BCD(E + e) + BCdE$ we have $A = BCD + BCdE$.

---

[46]Author's names.

[47]See [**103**, p. 62].

Jevons says "We may pass over Professor Boole's expression for A, after intrinsic elimination of $E(A = BCD + ABCd)$ as being sufficiently obvious.[48] By this he means strike out E by indirect inference, i.e., write A for E and we have $A = BCD + ABCd$.

We now require C in terms of A, B, and D, a problem considered by Boole on page 107 of *Laws of Thought*. Using Jevons's method of indirect inference, form all the possible combinations of A, B, C, D, E, and their contraries a, b, c, d, and e[49] and compare them with the premise, i.e., $A = BCD + BCdE$. Next consider all the combinations from $ABCde$ to $aBCdE$ which form a contradiction with one side of the premise. These can be discarded. The remaining terms make up the solution. Therefore, we proceed as follows:

Combining $ABCde$ with the premise $A = BCD + BcdE$, we have

$$AABCde = ABCBCDde + ABCBCddEe,$$

so

$$ABCde = ABCDde + ABCdEe$$

since AA=A, etc. from the Law of Simplicity. Therefore since $Dd = 0$ and $Ee = 0$ by the Law of Contradiction we have

$$ABCde = 0 + 0 = 0,$$

a contradiction with one side of the premise. This means that the expression $ABCde$ as a possible solution for C can be discarded, and other combinations are then considered and accepted or rejected by following the same method.

Selecting from the remaining the terms the containing C, we have

$$C = ABCDE + ABCDe + ABCdE + aBCde + abCDE + abCDe + abCdE + abCde.$$

Striking out the dual terms such as $E + e$, and intrinsically eliminating any remaining occurrences of E or e by substitution of C, we have

$$C = ABCD + ABCd + aBCd + abCd.$$

Eliminating C from $ABCD$, because $ABD = ABCD$, and striking out the dual terms $A + a$ and $D + d$, we have

$$C = ABC + aBCd + abC.$$

which can be interpreted as:

[48]*Ibid.* p 63.

[49]This notation was taken from his tutor, De Morgan.

> What is limited in supply is either wealth, transferable and either
> productive of pleasure or not, (*ABC*) or else some kind of what is
> not wealth, but is either not transferable (*abC*), or, if transferable, is
> not productive of pleasure (*aBCd*).

Jevons concluded that this is exactly equivalent to the solution obtained on page 108
of *Laws of Thought*

His own logical method is presented as superior to that of Boole; but because
his work is based to such a large extent on Boole's calculus, he finds himself in the
position of having to praise and condemn Boolean algebraic logic at the same time. In
PureLogic[50] he praised Boole's system with these words:

> Professor Boole's [system] is nearly or quite the most perfect system
> ever struck out by a single writer.

However he had previously remarked on the same page in putting forward his own
self-evident logic in comparison with Boole's "dark system":

> There are now two systems of notation, giving the same formal re-
> sults, one of which gives them with self-evident force and meaning,
> the other by dark and symbolic processes. The burden of proof is
> shifted, and it must be for the author or supporters of the dark sys-
> tem to show that it is in some way superior to the evident system.

Jevons adds, "It is not to be denied that Boole's system is consistent and perfect
within itself" but fettered, as Jevons believed with mathematical notation. In their
correspondence which took place before *Pure Logic* was published between 1863 and
1864, it is clear that Jevons was not afraid to state his objections to Boole's system. In
a letter sent in August 1863 he wrote bluntly:[51]

> I am desirous of early bringing to your notice objections that I have
> to urge against some of the principles of your logical system... Since
> I first became acquainted with your logical works some three years

---

[50]See [**103**, p. 67].
[51]See [**70**, p. 25].

since I have felt surprise at the apparent mixture of clearness and obscurity which your system presents.

3.4.3. Summary and Comparison with Boole's System. There are seven dimensions along which we can compare the systems of Jevons and Boole.

3.4.3.1. *Logical Addition.* Jevons differs from Boole in his definition of logical addition: classes no longer have to be distinct, and the definition of full union of intersecting classes is permitted. This move away from mathematical addition showed Jevons took a different position from Boole who had always stressed such analogies in his logic with mathematics. In fact the close relation between mathematics and logic that was perceived by Boole was probably the reason why he was so reluctant to give up his addition of disjoint classes. Jevons however had no such position and in fact wanted to discard such mathematical trappings.

3.4.3.2. *The Law of Unity.* The Law of Unity, $A + A = A$ was expressly ruled out by Boole, who held it to be uninterpretable, only giving as a consequence that A must be zero. Jevons challenged Boole to deny this law. Boole himself could be equally blunt. In a letter written on September 14, 1863, he replied:

> If I do not write more it is not from any unwillingness to discuss the subject with you but simply because if we differ on this fundamental point [the law of Unity] it is improbable that we should agree on others.

3.4.3.3. *Addition and Subtraction.* Because of the Law of Unity, Jevons asserts that addition and subtraction do not exist in thought or language generally but are only valid under a logical condition that logic imposes on number. His published correspondence with Boole includes a letter that he wrote on September 5, 1863:[52]

> My proposed alterations of your system however go further, for I altogether object to the use of the negative sign. I think that it has no place in logic, but is derived by arithmetic and maths from logic. ...I do not think that either addition or subtraction is a process of

---

[52]See [**70**, p. 28].

logic, but that the operations of logic consist in combination and separation of terms or notions.

3.4.3.4. *0/0, 1/0, 0/1 and 1/1.* Jevons objected to the symbols 0/0,1/0,0/1 and 1/1 as mathematical and mysterious, their interpretations only having been arrived at through study of particular examples.

3.4.3.5. *Indirect Inference.* Jevons's method of indirect inference is a long and tedious method of obtaining solutions to logical syllogisms in comparison with the more elegant method of Boole.

3.4.3.6. *Calculus of Form.* The logistic position of Jevons is revealed, in contrast to Boole's view that logic and mathematics are separate branches of a wider Calculus of Form. On page 71 of *Pure Logic*, in a chapter headed "Remarks on Boole's System, and on the Relation of Logic and Mathematics," Jevons stated his position:

> Number, then, and the science of number, arise out of logic, and the conditions of number are defined by logic.

3.4.3.7. *Completeness.* Jevons, like Peirce later, considered that Boole's algebraic logic was insufficient to express all propositions, e.g., Jevons used $A = AB$ for Boole's unsatisfactory $A = vB$ to express SOME B, and therefore could not explain the laws of reasoning. He went further and held that Boole's system contradicted the laws of thought, citing again Boole's insistence on an exclusive method of logical addition. In a letter to De Morgan on January 9, 1864, Jevons wrote:

> I think you will allow however that [Boole's] assumption of exclusive terms is quite contrary to the procedure of ordinary thought.

3.4.3.8. *Hypotheticals.* Jevons, as did Peirce, avoided Boole's hypothetical or secondary propositions which were propositions about propositions, i.e., whether such propositions were true or false. The effect of Jevons's *Pure Logic* was great although non-existent on Boole. Grattan-Guinness observes:[53]

---

[53]See [**70**, pp. 20–21].

... gradually Jevons's kind of approach became normal practise ... Under the influence of Jevons, ... Peirce, ... and other (partial) followers, Boole's algebraic approach to logic gradually grew in importance. However, several of Boole's key concepts and procedures were abandoned or changed: not only his reading of '+' but also his stress on (un)interpretability of logical functions and equations, his interest in solving logical equations, and notions such as 0/0 and 1/0 used to effect the solutions.

Boole's intractability also echoed his unwillingness even to read Jevons's *Pure Logic*. His two main excuses cited were overwork and the fact that he intended to publish on logic further and so wished to avoid any controversy over precedence of publication. He had avoided reading De Morgan's *Formal Logic* for the same reason. Such a fear of controversy was justified in view of the acrimonious dispute between the Scottish philosopher, Sir William Hamilton and De Morgan over who had precedence in publishing on the quantification of the predicate.[54]

Even though Boole could not agree with Jevons's objections to his system and in fact completely disagreed with his law of Unity and his use of addition between nondistinct classes he encouraged Jevons to publish in a short but concise note written on September 23, 1863:

> I beg leave to return to you with my best thanks Professor De Morgan's letter, and I have only to add that I entirely approve of the advice which he gives you believing that it is always to the ultimate interest of truth that objections to any particular system or doctrine should be stated by those who hold them, in the most unreserved manner.

Although Jevons was responsible for simplifying Boole's algebra greatly with his Law of Unity, $x + x = x$, the methods that he developed for his syllogistic problem solving were tedious. In particular his system of considering all possible combinations of his logical alphabet, i.e., all letters A, B, C, etc., and their contraries a, b, c, etc. It is also interesting to note that Jevons does not consider De Morgan's work on relations at all, instead preferring Boole's non-relative equations.

---

[54]See p. 59 above.

Like Jevons, Peirce can be seen in his next paper "On an Improvement to Boole's Calculus of Logic" to attempt his own modifications to Boole's work. He introduces new notation for logical operations to express the fact that the classes used are now no longer necessarily distinct. In this he follows the work of Jevons, but does not restrict himself to the operation of addition but also broadens his logic to include subtraction and division, operations expressly ruled out by Jevons in his more limited algebraic logic.

3.4.4. "On an Improvement in Boole's Calculus of Logic". In this paper it is clear that Peirce is now looking at a way of expressing relations between classes rather than expressing syllogistic logic. On page 12 he writes:

> Boole's Calculus of Logic ...consists, essentially of a system of signs to denote the logical relation of classes.

Peirce's modifications to Boole's notation include new symbols for equality and addition, viz., an equal sign followed by a comma $(=,)$ for the equal sign alone $(=)$ and a plus followed by a comma $(+,)$ for just the plus alone $(+)$.

In particular, for the definition of addition, the classes are not taken as disjoint. He states on page 12:

> Let $a + b$ denote all the individuals contained under a and b together. The operation here performed will differ from arithmetical addition in two respects: 1st, that it has reference to identity, not to equality; and 2nd, that what is common to a and b is not taken into account twice over, as it would be in arithmetic.

He then continues:

> If No a is b,

(1)
$$a + b = a + b$$

> it is plain that

(2)
$$a + a = a$$

> $+$ is then shown to be both commutative and associative.

Equation (2) is almost identical to the Law of Unity which, as we have seen, first appeared in Jevons' *Pure Logic*. Was Peirce influenced by Jevons's work, or did he discover this law independently? Peirce certainly knew of Jevons at the time this paper was written, as six months later on November 12, 1867, he presented a further paper to the Academy entitled "Upon Logical Comprehension and Extension" where he compared the different senses in which the terms *comprehension* (Hamilton's *intension*), and *extension* had been accepted by philosophers and logicians.[55] Peirce wrote in the November paper that Jevons applied the terms *comprehension* and *extension* to meanings. This is made quite clear in Jevons's introduction to *Pure Logic*[56], when he wrote:

> The number of individuals denoted forms the breadth or extent of the meaning of the terms; the qualities or attributes connoted form the depth, comprehension, or intent, of the meaning of the term. The extent and intent of meaning, however are closely related.

Peirce must have been aware of this viewpoint and also shared the belief of many English logicians that class extension was to be regarded as the sum of real external attributes. Continuing with the modifications in notation, Peirce uses a, b for ab in logical multiplication and defines this as:[57]

> Let a, b denote the individuals contained at once under classes a and b.

This operation is then shown to be commutative, associative and distributive with respect to addition. We have a minus followed by a comma (-,) for a lone minus (-) and two expressions separated by a semi-colon (a;b) for two expressions separated by a slash.

This represents not only a change of notation but a new logical operation, since Peirce's +, no longer involves disjoint classes, $a-, b$ is not completely, determinate. The new operation of +, can be seen to following closely Jevons' modifications of Boole's calculus, discussed previously in *Pure Logic* but as we have seen, Jevons ruled

---

[55]*Extension* is used here in an analogous sense to today's union of sets and comprehension is analogous to intersection of sets.

[56]See [**103**, p. 3].

[57]See [**144**, p. 13].

out the operations of subtraction and division, and so Peirce is forging a distinct and novel path with these two logical operations.

Unity or 1 is still not a Universe of Discourse but it is at least now defined in class terms.

Then unity denotes the class of which any class is a part [**144**, p. 15].

Peirce realized that because of the indeterminate results of his logical subtraction and division any interpretations of this logic are difficult to obtain.

The rules for the transformation of expressions involving logical subtraction and division would be very complicated.

So his method is to use logical addition and multiplication only to obtain the Duality Law,

$$x, (1 - x) = 0$$

and Development Theorem,

$$\varphi(x) = \varphi(1), x + \varphi(0)(1 - x).$$

Addition here involves disjoint classes, so Peirce uses + instead of +,. Furthermore,

-' is seen as the minimum of -

and

: is seen as the maximum of ;

The main difference to be seen in this paper is that Peirce has moved away from Boole's system to include new notation, retaining his operation of logical division which is absent from Boole's calculus and incorporating the concept that logical addition and subtraction are now an operations between classes that are not necessarily distinct. He expressed this on page 18 as:

$$a + b = (a+, b) + a, b.$$

The advantages of these modifications are given on page 21 of [**144**], in particular highlighting the problem that Boole faced with his indeterminate class v.

> The advantages obtained by the introduction of [logical addition and
> subtraction] are three, viz. they give unity to the system; they greatly
> abbreviate the labor of working with it; and they enable us to express
> particular propositions. This last point requires illustration. Let i be
> a class only determined to be such that only some one individual of
> the class a comes under it. Then $a-, i, a$ is the expression for SOME
> a. Boole cannot properly express SOME a.

In his next relevant paper to the American Academy of Arts and Sciences Peirce expanded his modifications. In "Upon the Logic of Mathematics" [**146**] instead of emphasizing the relations between classes without then going on to fulfil this promising
start, seems to begin by reverting back to traditional syllogistic logic. On page 59
Peirce wrote:

> The object of the present paper is to show that there are certain gen-
> eral propositions from which the truths of mathematics follow syllo-
> gistically ...

After initial definitions, Peirce then stated a number of theorems mainly showing commutativity or transitivity for the logical operations however as Peirce commented on
page 61:

> ... the proofs of most of which are omitted on account of their ease
> and want of interest.

The last of these theorems is Theorem XIV:

$$a+, a =, a.$$

Now the equivalent proposition $a + a = a$ was definitely ruled out by Boole because
his logical addition could only take place for distinct classes. $a + a = a$ for Boole
means that a must be the zero class. This theorem shows how far Peirce has moved
from Boole and how close he is to the position of Jevons. Later on page 69 in a section
headed "On Arithmetic," Peirce showed himself aligned to De Morgan in considering
the equality relation. For Boole = meant equivalence as in the sense of identity. But
Peirce made clear that at this point it is useful also to consider = in terms of equicardi-
nality. In a footnote on page 69 he wrote:

Thus, in one point of view, identity is a species of equality, and, in another, the reverse is the case. This is because the Being of the copula may be considered on the one hand (with De Morgan) as a special description of 'inconvertible, transitive relation,' while, on the other hand, all relation may be considered as a special determination of being.

Peirce ended the paper by reiterating that the laws of the Boolean calculus are identical with the laws of arithmetic for zero and unity; a point already discussed under the main distinctions of Boole's algebra and arithmetic. Peirce finishes:

These considerations ... will, I hope, put the relations of logic and arithmetic in a somewhat clearer light than heretofore.

3.4.5. "The Logic of Relations: Note 4". It is in Note 4 of "The Logic of Relations" [**147**, p. 88] that Peirce first introduced relative terms as a way of expressing particular propositions:

The mode proposed for the expression of particular propositions is weak. What is really wanted is something much more fundamental. Another idea has since occurred to me which I have never worked out but which I can here briefly explain.

If w denotes WISE, and s denotes SOLOMON, then the expression w^s cannot be interpreted by ... any principle of Boole's calculus. It might then be used to denote WISER THAN SOLOMON. Thus, relative terms would be brought into the domain of the calculus.

By analogy with algebra, Peirce obtains the formula

$$w^{(a+.b)} = w^a, w^b$$

or those who are wiser than the class of a and b are those who are both WISER THAN a and WISER THAN b. By using his operation of =, which is a logical operation on classes not relatives, Peirce is clearly reading these relative terms as relations.

In developing this exponential theory of relatives, however it soon becomes apparent that this form of expression is not possible without involved and convoluted notation. For example after specifying that n^s denotes NOT SOLOMON, Peirce then shows that to express the relative WISER THAN SOME MAN the term

$$n^{(n^w)^m}$$

is necessary. As before, even though he starts off with relative terms, he quickly turns his attention from n^m or NOT MAN, the relative term, to a relation showing that 0^m stands for NON-MAN. Thus deriving from a relative term, a relation. This move is a vital one, because having receiving the initial impetus from De Morgan's relatives, Peirce is still tied to the traditional syllogistic logic of Boole which deals with classes and therefore relations rather than relative terms, thus explaining the advantage in using relational terms wherever possible.

Peirce used his notation for relational terms such as 0^m to argue the traditional syllogism

> Let m be MAN and a be ANIMAL.
> Then EVERY MAN IS AN ANIMAL

as follows:

$$m =, m, a \text{ or } m, 0^a =, 0 \text{ or } m+, a =, a.$$

Then,

$$0^a =, 0^{m+,a} =, 0^m, 0^a \text{ or } 0^m =, 0^a+, 0^m.$$

Next Peirce introduced composition of relatives by letting h denote head,

$$\left(0^h\right)^m =, \left(0^h\right)^a +, \left(0^h\right)^m$$

and then

$$0^{(0h)m} =, 0^{(0h)a+,(0h)m} =, 0^{(0h)a}, 0^{(0h)m}.$$
$$0^{\left(0^h\right)^m} =, 0^{\left(0^h\right)^a+,\left(0^h\right)^m} =, 0^{\left(0^h\right)^a,\left(0^h\right)^m}.$$

That is, ANY MAN'S HEAD IS AN ANIMAL'S HEAD. Peirce here recognized the importance of the composition of relative terms. He stated:

> This result cannot be reached by any ordinary forms of Logic or by Boole's Calculus.[58]

---

[58]For a more detailed account of this argument see [**122**, p. 273]. Merrill notes that this is an instance of Peirce applying relations to classes rather than to other relations.

In this paper, Peirce also introduced subscripts as well as superscripts, mainly to denote the identity and inverse relations in the sense that if k denotes KILLER, then

$$k^m \quad \text{is KILLER OF EVERY } m$$
$$k^{k^m} \quad \text{is KILLER OF EVERY KILLER OF EVERY } m$$
$$k' \quad \text{is KILLER OF HIMSELF}$$
$$k_0^m \quad \text{is every } m$$
$$k_1^m \quad \text{is KILLED BY EVERY } m.$$

This discovery of the calculus of relations was one of the most important events in Peirce's logical development. His work on relations, which was taken up by Schröder and which influenced later logicians including Russell, ensures that Peirce is known today. Although it was De Morgan who originated the theory, Peirce developed and extended it. We shall see in the next chapter that De Morgan's work on logic and in particular on the logic of relations was of crucial importance to Peirce's own seminal paper "Description of a Notation for the Logic of Relatives" in 1870.

CHAPTER 4

Peirce's "Description of a Notation for the Logic of Relatives"

4.1. Introduction

Peirce's "Description of a Notation for the Logic of Relatives, resulting from an Amplification of the Conceptions of Boole's Calculus of Logic" [**149**] was communicated to the American Academy on January 26, 1870.[1] In this chapter, after a brief survey and summary of *Logic of Relatives* in Section 4.1, we investigate three main themes.

In Section 4.2, we identify the logical terms in *Logic of Relatives*, in order to clarify and classify them. Section 4.3 covers popular misconceptions arising from Peirce's treatment of absolute terms and classes and relative terms and relations. We will show that Peirce is sometimes confusing but not confused. In Section 4.4, we investigate the algebraic methods behind his mysterious treatment of differentiation and also provide examples of a logical interpretation of his differentiation thus demonstrating a sound logical as well as algebraic foundation to this process. Section 4.5 covers conclusions and comparisons with both Boolean algebra and De Morgan's logic of relations.

Peirce sets out his aims at the beginning of this paper. He intends, like his father in *Linear Associative Algebra*, only to provide a notation and language for his logic, leaving any applications or use to later scholars. *Logic of Relatives* comprises eighty pages and is divided into five main sections. The first section headed "General Definition Of

---

[1]This work will be referred to as *Logic of Relatives* henceforth. The pagination will be taken from Hartshorne and Weiss [**83**].

The Algebraic Signs" introduces the basic operations of inclusion[2], addition, multiplication, subtraction and division but mainly in terms of the associative, commutative and distributive laws. This algebraic approach is evident from the very beginning.

Definitions of the operations are not given but this is because they are either the standard invertible Boolean operations, or modifications made in earlier papers as we have seen in "Harvard Lecture VI," and in "On an Improvement in Boole's Calculus of Logic," both of which give non-invertible or non-determinate results. Two signs are used for each operation, e.g., x, y for commutative multiplication and xy for non-commutative multiplication, with respective operational inverses $x; y$ and $x : y$ for division. Similarly $x + y$ is the Boolean operation of addition and $x +, y$ is addition between classes not necessarily disjoint, as preferred by Peirce and Jevons. Peirce calls the former *invertible* addition and the latter *non-invertible,* meaning disjoint and non-disjoint respectively.

The operations of subtraction and division are only defined as the inverse functions of addition and multiplication given in the form of equations, e.g.,

$$(x - y) + y = x$$

although it is clear that subtraction operation $x - y$ is in the Boolean sense the class x with y removed from it, just as the Boolean addition operation is the taking-together operation. His definitions of algebraic operations are given in terms of algebraic equations. For example, he defines the zero term and unit term using the laws $x +, 0 = x$ and $x1 = x$. Peirce also plans to include Taylor's theorem which was essential to Boole's development theorem and therefore his whole algebraic logic, the transcendental number $\mathfrak{G}$ (or e which Peirce represents by the limit of the series $(1 + i)^{1/i}$) and the irrational number $\mathfrak{O}$ or π, 3.14159 as Peirce quotes it.[3]

In the second section, "Application of the Algebraic Signs To Logic," the three main kinds of logical terms – absolute, relative and conjugative – are introduced and the operations of multiplication and involution are developed. Peirce defines two forms of multiplication:

---

[2]Equality is defined in terms of inclusion as $x = y$ if $x \prec y$ and $y \prec x$, thus giving the inclusion relation, which Peirce terms illation, pre-eminence over the equality relation.

[3]Hartshorne and Weiss at [**83**, p. 33] point out that these symbols are identical to those used by Benjamin Peirce on page 11 of the lithographic version of *Linear Associative Algebra.*

a) functional or ordinary multiplication which is taken to be the concept of the application of a relation
b) logical or comma multiplication which introduces an extra correlate to the term.

Logical multiplication is commutative and therefore more useful to those seeking algebraic analogies, whereas functional multiplication is not commutative.

The operation of multiplication is now defined in terms of adjacent letters where $(sl)w = s(lw)$ as composition of relations, following De Morgan's "On the Syllogism: IV." Thus, (A SERVANT OF A LOVER OF) A WOMAN IS A SERVANT OF (A LOVER OF A WOMAN), so that the associative law is satisfied and the unit of this multiplication is defined to be the relative term 1 or IS IDENTICAL WITH _____, the first clear definition of the identity relation. Peirce has returned to De Morgan's notation rather than his own superscript notation for multiplication developed in "Note 4," although De Morgan used capital letters for relatives, e.g., LS for A LOVER OF A SERVANT OF, and, as Peirce comments on page 369, "he appears not to have had multiplication in his mind." On the other hand Peirce's other form of multiplication logical or comma multiplication l, s means WHATEVER IS A LOVER THAT IS A SERVANT OF _____ while functional multiplication ls means WHATEVER IS A LOVER OF A SERVANT OF _____.[4]

$1x = x$ defines the identity relation and a clear distinction is made between 1, i.e. unity or the universe or everything, and 1 the identity relation. Peirce also extends the Boolean concept of 1 or unity, in the sense of associating it with the number infinity. It is made clear that 1 is the universe and corresponds with infinity and that the symbol ∞ could be used for 1. In fact he does use this symbol for unity in a later logic paper[153]. Quantification is expressed by a superscript notation, e.g., l^w meaning WHATEVER IS A LOVER OF EVERY WOMAN.

The main bulk of theorems or formulae numbering from (1) to (85) is given in the next section, "General Formulae." In the first list of 33 general formulae, four are attributed to Jevons and ten to Boole. Although they use x and y as symbols of classes,

---

[4]The concept of multiplication here adopted by Peirce in 1870, is the application of one relation to another and on page 376 he observes "a quaternion being the relation of one vector to another, the multiplication of quaternions is the application of one such relation to a second." Earlier Peirce had also made the connection with quaternions when introducing involution. In a footnote on page 362 Peirce draws attention to the fact that Hamilton takes $(x^y)^z = x^{(zy)}$ instead of $(x^y)^z = x^{(yz)}$ a reference to the non-commutativity of quaternions.

it is clear from formula §7:

$$x, (y +\!\!, z) = x, y +\!\!, x, z$$

which is attributed to Jevons that the multiplication Peirce is using is that of multiplication of relatives:

WHATEVER IS AN x THAT IS A y OR A z IS AN x THAT IS A y OR AN x THAT IS A z.

Here the OR is used in the inclusive sense of EITHER OR BOTH.

He encompasses traditional syllogistic logic by including the principle of contradiction and excluded middle as formulae. The formula for principle of contradiction is

§25 $x, n^x = 0$

Here n stands for NOT. In fact n is the relative term NOT _____, and n^x is equivalent to $1 - x$. This notation was first seen in [**147**], when he first introduced relative terms as a way of expressing particular propositions. Later in *Logic of Relatives*, n is replaced by σ, probably motivated by the analogy with Newton's infinitesimal moment σx. The formula for the principle of excluded middle is

§26 $x +\!\!, n^x = 1.$

The final six propositions are stated to be derivable from the formulae already given, but no proofs are provided by Peirce. For example,

Proposition §28

$$(x +\!\!, y), (x +\!\!, z) = x +\!\!, y, z.$$

PROOF. First,

$$
\begin{aligned}
(xy), (x +\!\!, z) &= (x +\!\!, y)x +\!\!, (x +\!\!, y), z && \S7 \\
&= x, (x +\!\!, y) +\!\!, z, (x +\!\!, y) && \S9, x, y = y, x \\
&= x, x +\!\!, y, x +\!\!, x, z +\!\!, y, z && \S7 \\
&= x +\!\!, y, x +\!\!, x, z +\!\!, y, && \S23 \; x, x = x
\end{aligned}
$$

But
$$x +, y, x \text{ and } x +, x, z = x$$
and
$$x +, x = x$$
so from §22, x +, x=x, we have
$$(x +, y), (x +, z) = x +, y, z$$

The formulae §30–§33 involve a function φ which is commutative and are based on Boole's development theorem $\varphi x = \varphi 0 + \varphi 1 - (\varphi 0) x$. These are the only formulae for which sketchy proofs are given in a footnote.

The major applications of differentiation together with definitions of individual terms and infinitesimal and elementary relatives are developed in the fourth section, "General Method of Working with this Notation," which contains theorems §86 to §166. The last section headed "Properties Of Particular Relative Terms" contains the only example of syllogistic reasoning in *Logic of Relatives* and throws together disparate themes, some new e.g., simple relatives and converses or re-examines concepts such as conjugative terms. This section includes formulae §168–§172. A short conclusion at the end seems to have the effect of undermining the whole paper as Peirce here distinguishes from his theorems, those which cannot be derived from others and calls them axioms, only to repudiate them with the sentences:

> But these axioms are mere substitutes for definitions of the universal logical relations, and so far as these can be defined, all axioms may be dispensed with. The fundamental principles of formal logic are not properly axioms, but definitions and divisions.

The sections of the *Logic of Relatives* are as follows:

General Definition of the Algebraic Signs: The operations of addition, multiplication, subtraction and division are introduced.

Application of the Algebraic Signs to Logic: The logical terms are introduced. The operations of addition and multiplication are developed with two types of multiplication defined. The operation of involution (exponentiation) is introduced.

General Formulæ: Formulæ §1–§76 are listed and formulærelating to the numbers of terms are developed and listed in theorems §77–§85.

General Method of Working With this Notation: Theorems §86–§94 are developed and listed. Those formulae concerned with individual terms §95–§108 are obtained. Infinitesimal relatives are introduced as are the processes of differentiation and backwards involution. The appropriate theorems being §109–§153. The formulae developed from elementary relatives are listed in theorems §154–§166.

Properties of Particular Relative Terms: Here the classification of simple relatives, 'Not' or contrary terms, two examples of syllogistic reasoning, conjugative terms and converses are all investigated in the last set of theorems §167–§172.[5]

4.2. Peirce's Logical Terms

Under the heading "Application of the Algebraic Signs to Logic," Peirce defines three different kinds of logical terms:

a) *absolute* terms such as MAN, HORSE, TREE
b) *simple relative* terms such as FATHER OF, LOVER OF, SERVANT OF
c) *conjugative* terms such as GIVER TO _____ OF _____, or BUYER OF _____ FOR _____ FROM _____.

This three-fold categorization echoes his division in "On a New List of Categories" of logical categories into Quality, Relation and Representation.[6]

Peirce is careful to distinguish between absolute terms or classes and simple relative[7] and conjugative terms by the use of different typefaces, for example:

a ANIMAL   *e* {ENEMY OF _____}   g GIVER TO _____ OF _____

---

[5]Although Peirce numbered 172 theorems in *Logic of Relatives*, there are in fact 173 with §169 being used for two different theorems.

[6]See [**148**].

[7]Peirce uses 'relative' and 'relative term' interchangeably throughout *Logic of Relatives*.

It should be noted that Peirce is using a metatheoretical approach from the very beginning. He distinguishes between the terms which denote classes and the classes themselves. This reveals his great interest in semiotics.

In a later section on page 391 of *Logic of Relatives*, Peirce clarifies his meaning of individual terms as opposed to absolute terms, when he discusses the three types of logical terms that apply to individuals,

    a. An *individual term,* which denotes one specific individual.

    b. An *infinitesimal relative,* which is a relative term with the least number of correlates necessary for existence. If the number of its correlates is increased by one then no such relative terms exist. In particular a relative with a given number of individual correlates is an infinitesimal.

    c. An *elementary relative* which signifies a relation between mutually exclusive individuals or classes, or a relation between pairs of classes such that every individual of one class is in that relation to every individual of the other.

Infinitesimal and elementary relatives are instances of relatives. They can be either simple relatives or conjugative relatives because conjugative relatives are just made up of simple relatives combined as it were.

4.2.1. Absolute Terms and Individual Terms. Peirce firstly defines the individual term m as denoting ALL MEN.[8] So m represents the class of MEN or as Peirce writes later m denotes the class composed of MEN. Secondly, when individual terms are used with relative terms the meaning of the individual term changes to denote a member of the class, e.g., *l*w shall denote whatever is the LOVER OF A WOMAN.[9] Peirce is using his individual term A CERTAIN MAN as an individual representing every man or what Peirce refers to as an *individuum vagum* rather than the other type of individual term which refers to a specific individual, *individuum signatum,* such as JULIUS CAESAR.

> The *individuum vagum,* in the days when such conceptions were exactly investigated, occasioned great difficulty from its having a certain generality, being capable, apparently, of logical division. If we include under the *individuum vagum* such a term as 'any individual

---

[8]See [**157**, p. 368].
[9]See [**157**, p. 369].

man,' these difficulties appear in a strong light, for what is true of any individual man is true of all men. Such a term is in one sense not an individual term [individuum signatum]; for it represents every man. But it represents each man as capable of being denoted by a term which is individual; and so, though it is not itself an individual term [individuum signatum], it stands for any one of a class of individual terms.[10]

At the end of this passage Peirce goes on to say:

Thus, all the formal logical laws relating to individuals will hold good of such individuals by second intention, and at the same time a universal proposition may at any moment be substituted for a proposition about such an individual, for nothing can be predicated of such an individual which cannot be predicated of the whole class.

He is describing a way of deducing universally quantified conclusions.

Peirce calls his individual terms absolute terms. He writes an absolute term as a logical sum of individuals:

$$H = H +, H' +, H'' +, \ldots$$

Here $+,$ acts as the OR operator not the AND operator since these are individuals. Earlier on page 369 Peirce takes $l +, s$ to denote WHATEVER IS LOVER OR SERVANT OF _____.

Finally, individual terms when used with conjugative terms denote specific individuals, *individuum signatum*. He writes this on page 370 of *Logic of Relatives* as:

$$\mathfrak{g} \, o \, h = \mathfrak{g} \, o \, (H +, H' +, H'' +, etc.) = \mathfrak{g} \, o \, H +, \mathfrak{g} \, o \, H' +, \mathfrak{g} \, o \, H'' +, \ldots$$

So $\mathfrak{g} \, o \, h$ must be taken to mean whatever is the GIVER OF A HORSE TO THE OWNER OF THAT *horse* [Peirce's italics]. He emphasizes on page 371:

This is always very important. *A term multiplied by two relatives shows that THE SAME INDIVIDUAL is in the two relations.*

---

[10]See [**157**, p. 391ff.].

Some consideration must now be given to the way in which Peirce uses the word 'denote'. In *Logic of Relatives* he attempted to provide a syntactical structure which is concerned with notation rather than semantics. His terms are syntactical objects, e.g., the subject of a proposition is a term. A predicate of a proposition is a term. However Peirce speaks of terms as denoting classes:

> I propose to use the term 'universe' to denote that class of individuals about which alone the whole discourse is understood to run.[11]

Here he is trying to indicate why he constructs the notation and uses the word 'denote' informally in a semantic exposition. Although considerations of meaning, denotation and class are not involved at the syntactical level, Peirce uses the word 'denote' informally in the sense of 'represent' or 'stand for'. Burch [**21**, 208] has:

> And of course relative terms do denote classes, when the semantics is extensional in structure and when the sense of the word 'denote' is understood to be given some interpretation function which connects syntax with this semantics.

Peirce also defines the number associated with an absolute term. For example, he does not define [m] as the number of men in the class mankind but rather as the number of individuals denoted by a representative man. He writes "... the difference between [m] and [m,] must not be overlooked." [m] in fact stands for the number of individual men represented by the absolute term m, while [m,] is the average number of MEN THAT ARE _____. Peirce defines the number associated with a term, that is an absolute term, as the number of individuals it denotes, but for relative terms it is the average number of individuals that are related to one individual. For example, [*t*] where *t* stands for TOOTH OF then [*t*] is 32. So here the rank or cardinal number of a class of a relative term does not necessarily correspond to the number of individual relatives in the class. It would be difficult however, to decide upon [*e*] or the average number of enemies of one individual. Peirce uses average not in the sense of arithmetic mean but rather to indicate the normal, or standard rational number associated with the relative term.

4.2.2. Relative Terms and Conjugative Terms. When looking at relative terms in defining addition between the two relative terms *l* denoting WHATEVER IS A LOVER

---

[11] See [**157**, p. 366].

OF_____ and s denoting WHATEVER IS A SERVANT OF _____, Peirce uses $l +\!\!, s$ to denote WHATEVER IS LOVER OR SERVANT OF_____, i.e. the class of lovers or servants. So while absolute terms are treated as objects or instances of a class, relative terms (and conjugative terms) must be regarded as representing classes themselves. This is also seen on page 377 when introducing the binomial term $(e +\!\!, c)^f$ which denotes not the individual term AN EMPEROR OR CONQUEROR OF EVERY FRENCHMAN but as Peirce writes "those things each of which is emperor or conqueror of every Frenchman." So clearly relative terms represent classes rather than individuals. Peirce in the introduction to *Logic of Relatives* refers to De Morgan's notation $X..LY$ as signifying some object X is one of the Ls of Y. He himself represents the class of objects X by ly, the class of LOVERS OF y, where l is a relative term representing the class of lovers and y is an absolute term, a representative of a class.

This move away from the logic of classes as seen in Boole, to include the concept of individual or representative members of classes can be seen in De Morgan who used X and Y to denote individuals rather than classes, and L to denote a relation (in the singular rather than a class of relations). For example in "On the Syllogism: IV," page 222, he refers to LM to denote an L OF AN M. Peirce often uses the phrase 'a lover' instead of 'whatever is a lover', for the meaning of l following De Morgan's lead, although l signifies here the class of lovers.

To summarize, while De Morgan looks at relations, e.g., LOVES, denoted by L, between individual members of classes and composition of such relations, Peirce is interested in applying classes of relatives, e.g., WHATEVER IS A LOVER, denoted by l, to absolute terms. Such absolute terms while denoting classes are also individual representatives of classes as signified by lw or WHATEVER IS A LOVER OF A WOMAN. This is also reflected in the titles of the two works on relational theory. De Morgan's "On the Syllogism: IV" and on the "Logic of Relations" emphasizes relations whereas Peirce's title has "Logic of Relatives." The use of a relative and individual term ly is similar to the use of a relation and an individual term by De Morgan as in $..LY$. Peirce refers to this application of a relation as multiplication and he writes on page 369, "this notation is the same as that used by Mr. De Morgan, although he appears not to have had multiplication in his mind."

So Peirce's notion of the application of relation to a class lw is not essentially new, but formalized and extended to the application of a relation to a relation as relative multiplication. However he does not concern himself with relative composition as De Morgan did, and only briefly shows that the distributive law for addition and the associative law for multiplication hold:

$$(l +\!\!, s)\text{w} = l\text{w} +\!\!, s\text{w} \quad \text{and} \quad (sl)\text{w} = s(l\text{w}).$$

In his use of conjugative terms, because such terms as $\mathfrak{g}$ or WHATEVER IS A GIVER TO _____ OF _____ have a fixed order, Peirce has to abandon associativity and instead introduces a subscript notation in order to identify correlates, e.g., in $\mathfrak{g}_{12}l_1$wh,$\mathfrak{g}$ denotes GIVER TO_____ OF _____ and the first subscript 1 indicates that the first correlate is to be the first letter to the right of it, i.e. l and the second subscript 2 indicates that the second correlate is to be found two letters to the right of l. Since l is itself followed by 1 this means that its correlate is to be found one letter to the right of it, i.e. w, so that $\mathfrak{g}_{12}l_1$wh denotes GIVER OF A HORSE TO A LOVER OF A WOMAN. It can be seen that

$$\mathfrak{g}_{12}l_1\text{wh} = \mathfrak{g}_{11}l_2\text{hw}.$$

Also we have $\mathfrak{g}_{2\text{-}1}\text{h}l_1$w where the subscript 2-1 denotes that the first correlate is to be found two letters to the right of $\mathfrak{g}$, i.e. l and the second correlate is one letter to the left of l. l_0 denotes LOVER OF HIMSELF. It is interesting to remember that Peirce has used this subscript notation before in "Note 4"[12] but to express the identity relation so that l_0 would denote 1. To denote the reflexive relative, LOVER OF HIMSELF, he used a superscript prime l' rather than a subscript as here. In l_∞, ∞ stands for an indeterminate correlate so l_∞ denotes LOVER OF SOMETHING.

These subscript numbers for conjugative terms are difficult to work with, and Peirce himself says on page 372 "these subjacent numbers are frequently inconvenient in practice." He also has an alternative notation using reference marks so that

$$\mathfrak{g}_{\dagger\ddagger}{}^\dagger l_{\|}{}^{\|}\text{w}^\ddagger\text{h}$$

denotes GIVER OF A HORSE TO A LOVER OF A WOMAN. Corresponding reference marks link the terms to the appropriate correlate and moreover indicate which correlate is to be taken first. Since $\mathfrak{g}$ is the conjugative term indicating the class represented by GIVER TO _____ OF _____, the $\dagger$ mark indicates where the first correlate may be found. The eye is then drawn along the expression to the corresponding $^\dagger l$ so that GIVER TO A LOVER is obtained. The second reference mark $\ddagger$ read in order from left to right indicates the second correlate of $\mathfrak{g}$, and the eye is drawn to ‡h so that GIVER TO A LOVER OF A HORSE is obtained. The $\|$ refers to the correlate of l and matching this with $^\|$w gives LOVER OF A WOMAN. The final interpretation is

---

[12]See page 113.

WHATEVER IS A GIVER TO A LOVER OF A WOMAN, OF A HORSE.

With this notation conjugative terms are associative under multiplication and Peirce regards this as similar to relative multiplication or the application of a relation – the application of the conjugative term g to relative term and absolute term.

Peirce uses the term *functional multiplication* to refer to the application of relatives or conjugatives. He does not however, continue with either form of subjacent number or reference mark notation to any great extent in this paper, except in his version of the binomial theorem when he uses a mixture of both notations. He also shows by the following example that any conjugative term (defined earlier as a relative term having at least two correlates) having more than two correlates can be reduced to a combination of conjugatives of two correlates:

$$w = uv$$

where w stands for WINNER OVER OF_____, FROM _____, TO _____ and u is the conjugative term GAINER OF THE ADVANTAGE _____, TO _____ and the first correlate here is v or THE ADVANTAGE OF WINNING OVER OF _____, FROM_____.

This can be seen more clearly by introducing the absolute terms x, y and z. Taking v to be THE ADVANTAGE OF WINNING OVER OF x FROM y, and u to be GAINER OF THE ADVANTAGE _____ TO z; then uv is GAINER OF THE ADVANTAGE OF WINNING OVER OF x FROM y, TO z. Abbreviating GAINER OF THE ADVANTAGE OF WINNING to WINNER, uv is now equivalent to w. A more convenient notation is also introduced for the expression of conjugative terms rather than the subscript notation used earlier, i.e. $g_{12}l_1$wh denoting GIVER OF A HORSE TO A LOVER OF A WOMAN. Since conjugatives can be reduced to conjugatives of two correlates introducing a symbol between the two correlates would remove the need for such subscripts.

The associative law (although Peirce refers to this as the associative principle) for a conjugative relation is expressed as

$$x \mathbin{\mathsf{J}} (y \mathbin{\mathsf{J}} z) = (x \mathbin{\mathsf{J}} y) \mathbin{\mathsf{J}} z.$$

The conjugative term EITHER OR could then be represented by the symbol '+ which indicates an associative and commutative operation in the typical Peircean example on page 428:

If p denotes PROTESTANTISM, r ROMANISM, and f WHAT IS FALSE, then
p '+ r≺f means EITHER ALL PROTESTANTISM OR ALL ROMANISM IS FALSE.
In this way it is plain that all hypothetical propositions may be expressed.

This conjugative term '+ is therefore analogous to the exclusive or operation in today's
set theory.

In fact Peirce's three main categories of terms, i.e. absolute terms, relatives and
conjugatives are connected by an operation – the operation of logical multiplication.
This operation is introduced by Peirce for the first time in *Logic of Relatives* as an alternative to relative or functional multiplication. Functional multiplication is represented
by the juxtaposition of terms and as shown in Section 4.1 consists of the application
of a relation but it is not however commutative. Now Peirce introduces a new type
of multiplication called *comma* or logical multiplication which is commutative. This
multiplication has the effect of converting absolute terms into relative terms by the addition of an extra correlate, e.g., it transforms the absolute term m to m, the relative
term MAN THAT IS. Peirce says:

> I shall write a comma after any absolute term to show that it is so
> regarded as a relative term . . . any relative term may in the same way
> be regarded as a relative with one correlate more.

In other words, it transforms the relative term *l* or LOVER OF _____ to the term *l*, or
LOVER OF _____ THAT IS _____. It is clear that in this way relative terms are transformed
into conjugative ones.

Peirce does not state this explicitly but does compare m,,b,r with ɡoh and indicates
they should be interpreted in a similar way. Here in m,, the absolute term m is transformed into a conjugative term m,, meaning MAN THAT IS _____, THAT IS _____ by two
applications of logical multiplication and can be compared with the conjugative term
ɡ GIVER TO _____ OF _____ which also has two correlates. The term b, or BLACK INDI-
VIDUAL THAT IS _____ can be compared with the relative term *o* or OWNER OF _____,
while r or RICH INDIVIDUAL can be compared with the absolute term h a HORSE.

Let us consider the term m,,b,r, which is interpreted as

A MAN THAT IS A RICH INDIVIDUAL AND IS A BLACK THAT IS THAT RICH INDIVIDUAL.

Two things should be noted. Firstly r should be taken as the first correlate and b, r as the second correlate. Secondly, r represents a specific rich individual, THAT RICH INDIVIDUAL, rather than a representative individual, which is also the correlate for the relative term b,. So for example, taking the specified rich individual as John, we have A MAN THAT IS JOHN IS THE BLACK INDIVIDUAL THAT IS JOHN. This corresponds to the conjugative term g𝑜h denoting GIVER OF BLACK BEAUTY TO THE OWNER OF BLACK BEAUTY where Black Beauty is the specific individual horse h.

Peirce then writes some apparently contradictory sentences. He states that:

> ... after one comma is added, the addition of another does not change the meaning at all, so that whatever has one comma after it must be regarded as having an infinite number.

However he then provides a counter-example with a none too satisfactory explanation.

> If, therefore, $l,,sw$ is not the same as l,sw (as it plainly is not, because the latter means a LOVER AND SERVANT OF A WOMAN, and the former a LOVER OF AND SERVANT OF AND SAME AS A WOMAN), this is simply because the writing of the comma alters the arrangement of the correlates.

Peirce is saying that although the meaning may not change the syntactic rule for applying it may change.

To clarify these statements, we have to note that an absolute term is transformed by logical multiplication (represented by a ',') into a relative term so that it has one correlate. The repeated application of the comma transforms it to a conjugative term and identifies it with the correlate, which as we have seen with conjugative terms is a specific individual. Further repetitions continue the identification, i.e. THAT IS JOHN, THAT IS JOHN, etc. so that the meaning is unchanged. However for a relative term transformed into a conjugative term l, LOVER OF _____ THAT IS _____ unless the final correlate, i.e. the correlate of the final 'THAT IS is an absolute term the meaning of the expression will be changed. In the example given above, l,sw the final correlate is sw or 'servant of a woman'. The meaning of l,sw here is WHATEVER IS A LOVER OF A WOMAN

THAT IS A SERVANT OF A WOMAN. Since the correlate sw is not an absolute term, $l, , sw$ has a different meaning, i.e. LOVER OF A WOMAN THAT IS A SERVANT OF THAT WOMAN THAT IS THAT WOMAN. Now since the final correlate is an absolute term, i.e. w it can be seen that repeated logical multiplication will not change the meaning but only repeat THAT IS THAT WOMAN.

4.2.3. Individual Terms and Elementary Relatives.
Peirce's individual terms denote only individuals. Martin has noted the similarity between individual terms and unit classes, i.e. classes with a single member.[13] In fact within the part/whole class theory of Boole and De Morgan this was the only means of representing individual terms. The two fundamental formulae relating to individuality are given by §95 and §96.

(§95) $$x > 0 \Rightarrow x = X +, X' +, X'' +, \ldots$$

(§96) $$y^X = yX.$$

Here the individuals are denoted by capitals.

Peirce now proceeds to prove §102, $s^W \prec sw$ when w does not vanish.

Here is the proof:

$$
\begin{aligned}
s^W &= s^{W' +, W'' +,} \ldots && \text{§95} \\
    &= s^{W'}, s^{W''}, \ldots && \text{§11} \\
    &= sW', sW'', \ldots && \text{§96} \\
    &\prec sW', sW'', \ldots && \text{§96, §21} \\
    &= s(W' +, W'' +, W''' \ldots) && \text{§5}^{14} \\
    &= sw && \text{§95}
\end{aligned}
$$

Peirce introduces elementary relative terms on page 408 and defines them as relations between mutually exclusive pairs of individuals, or as relations "between pairs of classes in such a way that every individual of one class of the pair is in that relation to every individual of the other." All relatives then can be expressed as a logical sum

---

[13]See [**121**, p. 31].

[14]Peirce has (31) here, which is a formula relating to the development theorem of Boole and involves a function of one variable, rather than (21). He does not mention §5, §11 or §94.

of elementary relatives.[15] These relatives are defined in terms of the classes using the following procedure. $A : B$ denotes the elementary relative which multiplied into B gives A, e.g., TEACHER:PUPIL. Here we assume that every teacher teaches every pupil in a school.

$A : B$ gives rise to a system of four elementary relatives of the form $A : A$, $A : B$, $B : A$, and $B : B$. Taking A as TEACHER and B as PUPIL we have the elementary relatives, COLLEAGUE:TEACHER, TEACHER:PUPIL, PUPIL:TEACHER, and PUPIL:SCHOOL-MATE.

The mutually exclusive classes BODY OF TEACHERS IN A SCHOOL and BODY OF PUPILS IN A SCHOOL are called the *universal extremes* of that system. A common characteristic between the individuals of a pair is called a scalar and we have f, being a scalar supposing FRENCH TEACHERS have only FRENCH PUPILS, and vice versa. This is seen in

(§154) $s, r = rs$

where r is an elementary relative and s is a scalar. A logical quaternion then is defined as any relative which is capable of being expressed in the form

$$a, c + b, t + c, p + d, s$$

where c, t, p, and s are the four elementary relatives of any system and a, b, c, and d, are scalars representing COLLEAGUE OF, TEACHER OF, PUPIL OF, and SCHOOLMATE OF. Since any relative can be expressed as the logical sum of elementary relatives, so any relative may be regarded as resolvable into a logical sum of logical quaternions.

If u and v are the universal extremes of the system, i.e. the mutually exclusive classes, then we may write

$$c = \text{u:u}t = \text{u:v}p = \text{v:u}s = \text{v:v}.$$

Here each relative is defined in terms of a pair of mutually exclusive classes, and although Peirce calls this an elementary relative, he is also defining a relation between the two classes.

We have

$$(w'; w)\, \sigma^{-w} = 0$$

---

[15]Martin is mistaken when he quotes Peirce as writing "every relation may be conceived of as a logical sum of elementary relatives" [121, p. 19]. In fact Peirce writes that every *relative* may be so conceived, and elementary relatives are not relations which exist between mutually exclusive pairs etc. but *signify* such relations.

where w′ and w can be any of u or v, i.e. THE TEACHER OF ANY PERSON NOT A PUPIL is 0, and σ^- represents the relative term NOT _____ . The multiplication table of a system of elementary relatives is given in the table below.

(§156)

	c	t	p	s
c	c	t	0	0
t	0	0	c	t
p	p	s	0	0
s	0	0	p	s

This is the four by four multiplication table in Benjamin Peirce's *Linear Associative Algebra*. Peirce lists all sixteen of the propositions expressed by this table in ordinary English, e.g. on page 410:

> The colleagues of the colleagues of any person are that person's colleagues;
> The colleagues of the teachers of any person are that person's teachers; etc.

All relatives then can also be expressed in the form $a,i + b,j + c,k + d,l + \ldots$ where i, j, k, l, etc. are logical quaternions. Then the multiplication table for i, j, k, l, $\ldots$ provides a method of finding the multiplication of relations between more than two classes of individuals. For example the multiplication table involving three classes of individuals u_1, u_2, and u_3 involve nine logical quaternions. Peirce provides the equivalent 9×9 multiplication table and by certain substitutions, i.e.

$$i = u_1 : u_2 + u_2 : u_3 + u_3 : u_4, \qquad j = u_1 : u_3 + u_2 u_4, \qquad k = u_1 : u_4$$

and obtains the following table:

	i	j	k
i	j	k	0
j	k	0	0
k	0	0	0

This table holds since $(u_p : u_q)(u_r : u_s) = 0$ unless $q = r$ in which case the result is $u_p : u_s$, so that for example,

$$ii = (u_1 : u_2 + u_2 : u_3 + u_3 : u_4)(u_1 : u_2 + u_2 : u_3 + u_3 : u_4)$$
$$= 0 + u_1 : u_3 + 0 + 0 + 0 + u2 : u_4 + 0 + 0 + 0$$
$$= j.$$

It also appears as (b_3) on page 51 of the lithographic version of *Linear Associative Algebra*. Similarly another substitution for i, j, k, l and m gives a 5×5 table.

Although Charles Peirce at this point writes on page 413 "these multiplication-tables have been copied from Professor Peirce's monograph on Linear Associative Algebras;" in fact the 9×9 table is not in *Linear Associative Algebra* since Benjamin Peirce did not proceed beyond the sextuple algebras and the 5×5 table given by Charles is also not included in the quintuple algebras in *Linear Associative Algebra*. Charles continues on page 413:

> I can assert, upon reasonable inductive evidence, that all such alge-
> bras can be interpreted on the principles of the present notation in
> the same way as those given above. In other words, all such alge-
> bras are complications and modifications of the algebra of §156. It
> is very likely that this is true of all algebras whatever.

He was later to prove that all linear associative algebras could be expressed in terms of elementary relatives[16], and again in the addenda of *Linear Associative Algebra* published in the *American Journal of Mathematics*.[17] Merrill adds in his introduction to volume two:

> This technique formed the foundation of the method of linear repre-
> sentation of matrices, which is now part of the standard treatment of
> the subject.[18]

---

[16]See [150].
[17]See [153].
[18]See [83, p. xlv].

Charles also states that the algebra of §156 has been shown by Benjamin Peirce to be the algebra of Hamilton's quaternions. He then defines 1, i, j, k as the four fundamental factors of quaternions in terms of scalars a, b, c and $J = \sqrt{-1}$. The 4×4 resulting multiplication table is given. He claimed such tables can be used to produce logical interpretations for his logic of relations by firstly transforming the given logical expression into a linear combination of elementary relatives which can then be represented in the form of Hamilton's quaternions and then making use of geometrical reasoning.

For example, consider the quaternion relative

$$q = xi + yj + zk + wl$$

where x, y, z, and w are scalars. q can have the geometrical interpretation of a scalar, vector, versor, tensor or conjugate if it takes one of the forms given in formulae §157–§164, e.g.,

$x(i + 1)$	§157, scalar
$xi + yj + zk - xl$	§158, vector
$xi + xyj + \frac{z}{yk} + zl$	§160, zero
$Sq = \frac{1}{2}(x + w)(i + 1)$	§161, scalar of q

Chris Brink notes that this is an attempt by Peirce to contribute to the foundations of geometry via the notion of a logical quaternion.[19] In order to exhibit the logical interpretations of these functions, on page 415 Peirce considers a universe of married monogamists in which HUSBAND and WIFE always have COUNTRY, RACE, WEALTH, and VIRTUE[20] in common. Let

i denote MAN THAT IS _____
j denote HUSBAND OF_____
k denote WIFE OF _____
l denote WOMAN THAT IS _____

---

[19]See [**18**, p. 285].
[20]That is to say that they are virtuous if they are not thieves.

x denote BLACK THAT IS _____
y denote RICH PERSON THAT IS _____
z denote AMERICAN THAT IS _____
w denote THIEF THAT IS _____

q then as defined previously, is the class of

BLACK MEN, WOMEN-THIEVES, RICH HUSBANDS AND AMERICAN WIVES

$2Sq$ denotes

ALL BLACKS AND ALL THIEVES

However this clearly follows from §161 above rather than §160 as quoted by Peirce.[21] He goes on to explain that geometry can then be applied to the logic of relatives and gives the example that Euclid's axiom concerning parallels corresponds to the quaternion principle that the square of a vector is a scalar. He continues:

> From this it follows, since by §157 $yj + zk$ is a vector, that the rich husbands and American wives of the rich husbands and American wives of any class of persons are wholly contained under that class, and can be described without any discrimination of sex. In point of fact, by §156, the rich husbands and American wives of the rich husbands and American wives of any class of persons, are the rich Americans of that class.

However it is quite clear that §156 as used above refers to §157, since only two pages previously on page 413, Peirce was referring to §156 as the standard 4×4 table for the multiplication of quaternions. Here his numbering is consistently one out, as §157 refers to §158.[22] The vector $yj+zk$ which represents RICH HUSBANDS AND AMERICAN WIVES when composed with itself should give a scalar and the form of a scalar as given in §157 is a logical expression in which the relative applies to both MEN and WOMEN, i.e. the relative is independent of sex, which leads to the deduction made by Peirce.

---

[21]Also noted by Hartshorne and Weiss in [**84**, p. 82].

[22]This is possibly due to the fact that two versions of the binomial theorem are given in theorem §12 which were originally §12 and §13.

He does not pursue the geometrical analogy for much longer. He writes on page 417:

> It follows from what has been said that for every proposition in geometry there is a proposition in the pure logic of relatives. But the method of working with logical algebra which is founded on this principle seems to be of little use. On the other hand, the fact promises to throw some light upon the philosophy of space.

4.3. Misconceptions and Errors in Peirce's *Logic of Relatives*

4.3.1. The Confusion between Absolute Terms and Classes. Some theorems of Peirce which need further analysis to clarify the confusion between what is a class and what is an absolute term, will now be highlighted with additional comments. Peirce on page 392 in the section headed *Individual Terms*, uses formulæ

§94: $x', x'' \prec x'$
§21: $x' \prec x' +, x''$
§1: If $x \prec y$ and $y \prec z$, then $x \prec z$,

to get
$$x', x'' \prec x' +, x''$$
and so produce the formula
$$x', x'', x''', \ldots \prec x' +, x'' +, x''' +, \ldots$$
which he writes as §101,
$$x', x'', x''', etc. +, \ldots$$
or §101
$$\Pi' \prec \Sigma'$$

where Π' and Σ' signify that the multiplication and addition with commas are to be used.[23] Putting $s = ls$ in §102, $s^w \prec sw$, we have
$$(ls)^w \prec lsw$$

---

[23]Confusingly Peirce writes "addition and multiplication" here.

Also since by §94, $a, b \prec a$, we have

$$la, b \prec la.$$

Similarly

$$la, b \prec lb.$$

Multiplying,

$$la, b \prec la, lb$$

§106 is stated as

$$l^{s^w} \prec ls^w$$

This also follows directly from §102 by taking $a = l^s$, $b = ls$ and $c = $ w in §93, $a \prec b \Rightarrow ac \prec bc$.

Peirce then proceeds to prove §107,

$$l^s w \prec l^{s^w}.$$

Martin writes:

> The next formula is supposed to express EVERY LOVER OF EVERY SER-
> VANT OF ANY PARTICULAR WOMAN IS A LOVER OF EVERY SERVANT OF ALL
> WOMEN, §107
>
> $$l^s W \prec l^{s^w}$$
>
> Here W is understood to be a particular WOMAN. But clearly the
> verbal reading is false, and likewise the formulae. On the other hand,
> its converse is true. EVERY LOVER OF EVERY SERVANT OF ALL WOMEN IS A
> LOVER OF EVERY SERVANT OF ANY PARTICULAR WOMAN, so that §108
>
> $$l^{sw} \prec l^s W$$

In fact, Peirce states §107 as $l^s w \prec l^{s^w}$ and §108 as $l^{sw} \prec l^s w$ with the lower case w
not W. Martin is mistaken here in supposing that w stands for a particular, individual
WOMAN, i.e. an individual term. But w is an absolute term and

$$w = W' +, W'' +, W''' +, \ldots$$

which is a representative member of the class of WOMEN. Since it is true that A SERVANT
OF ALL WOMEN IS A SERVANT OF A (SOME) WOMAN SO, verbally it is true that EVERY LOVER
OF EVERY (SERVANT OF A WOMAN) IS A LOVER OF EVERY (SERVANT OF ALL MEN), provided A
SERVANT OF ALL WOMEN exists. This follows from §92, $a \prec b \Rightarrow c^b \prec c^a$. Compare this

with the stronger $l^{sw} \prec l^{s^w}$ which Peirce did not prove but follows directly from §102 and §92.

Formula §108 given by Peirce and quoted by Martin is not of course the converse of §107. The converse is $l^{s^w} \prec l^s w$, which does not following directly from §108 since we have $l^{sw} \prec l^{s^w}$ rather than $l^{s^w} \prec l^{sw}$. Martin also writes:[24]

> Let ɢ stand for the triadic relation of giving. Then Peirce lets ɢxy stand for the class of GIVERS OF Y TO X ... But note here that x and y are no longer classes but individuals.

Not only is confusion generated by uncertainty over the exact use of absolute terms which could have been easily clarified by Peirce but also there seems to be a confusion over the distinction between absolute terms and relative terms. This arises from the fact that Peirce often uses absolute terms instead of relative terms in certain theorems concerning relative terms. Formulae §102–§108, §126–§132 and §145–§147 contain absolute terms. However, it will become apparent that Peirce viewed absolute terms as interchangeable with relative terms through his process of logical multiplication. Merrill writes:

> For all its importance, the *Logic of Relatives* memoir presents many problems of interpretation. Perhaps the most serious issue is whether Peirce is dealing with relations or with relatives – that is with the relation of being a servant, or with such classes as the class of servants ... but in some cases his terms clearly stand for relations. The situation is complicated by the fact that many terms, such as 'servant' can stand for either a relation or a relative, depending upon the context.[25]

But as we have seen, in the discussion of his logical terms, Peirce does not deal with relations per se as De Morgan did, only in relative terms denoting such relations. The confusion here is that SERVANT is taken to stand for the relation SERVES but is in fact an absolute term. It is the case that the confusion arises between absolute terms and relative terms, not between relations and relative terms. In order to clarify Peirce's position

---

[24]See [**121**, p. 27].
[25]See [**83**, p. xlv].

regarding absolute terms and relative terms and also how their use is interchangeable it will be of use to first investigate the operation of multiplication.

Multiplication between relatives is taken by Peirce to mean composition of relations. He stated on page 372 "our conception of multiplication is the application of a relation" and he defined multiplication between absolute terms in the following way. He treated the absolute term MAN as a relative term MAN THAT IS _____ denoted by m,. In this way, an absolute term (m) is transformed into a relative term (m,). With more than one relative, the term immediately to the right of the comma is taken as the correlate of 'THAT IS _____' i.e. l, sw denotes A LOVER OF A WOMAN THAT IS THE SERVANT OF THAT WOMAN. In the same way applying the comma twice, we have m,,b,r or A MAN THAT IS A RICH INDIVIDUAL AND IS A BLACK THAT IS THAT INDIVIDUAL, where r is interpreted as RICH INDIVIDUAL. This is the same as m,b,r or MAN THAT IS BLACK THAT IS RICH, so the application of a comma alters the order in which the correlates are taken.

It is possible to write l, sw= l, sw, 1, 1, 1, 1, 1, ... where 1 denotes w each time or as Peirce commented, "all these ones denote the same identical individual denoted by w." This is further evidence that Peirce regarded the absolute term w when used with relatives as an object or representative member of a class so that a term may be regarded as having an infinite number of factors, those at the end being the identity relative 1, signifying IDENTICAL WITH _____. This comma multiplication Peirce refers to as logical multiplication as opposed to relative or functional multiplication. Logical multiplication unlike functional multiplication is commutative with identity 1_0 or 1, and as Peirce commented on page 375 of *Logic of Relatives*, is effectively the same as that of Boole in his logical calculus. Brink has noted the analogy with set-theoretic intersection of classes.[26]

Peirce's operations such as logical product, denoted by a ',' and ordinary or functional multiplication, that is, composition of relatives, are only defined between classes as in the Boolean logic of classes; but occasionally as in formula §168 he uses absolute terms and relatives interchangeably. This, as we have seen, he was able to do by using the relative term w, for the absolute term w meaning a woman. w, is then WHATEVER IS A WOMAN THAT IS _____ and is a relative rather than an absolute term. So these formulæ are equally valid for absolute terms as for relative terms. Peirce wrote on page 372 of *Logic of Relatives*:

---

[26]See [**18**, p. 289].

Now the absolute term "man" is really exactly equivalent to the relative term "man that is _____," and so with any other. I shall write
a comma after any absolute term to show that it is so regarded as a
relative term.

And again on page 373 Peirce added "I shall write a comma after any absolute
term to show that it is so regarded as a relative term ... any absolute term [may] be thus
regarded as a relative term." This interchangeable use of an absolute term for a relative
term has led to the opinion that Peirce confused relations and relative terms. Lewis
[**115**, p.95] has: "Peirce, like De Morgan, treats his relatives as denoting ambiguously
either the relations or relative terms. *a* is either the relation "agent of" or the class
name "agent.""

Here of course it is Lewis who is confusing the relation IS THE AGENT OF with the
relative term AGENT OF and the class name AGENT for the absolute term AGENT. The fact
that Peirce regarded absolute terms and relative terms as interchangeable is not to say
that he confused the two concepts. On the contrary, on page 366, he introduced all the
letters that he used together with their meaning and groups them according to whether
they are absolute, relative or conjugative.

In *Logic of Relatives* Peirce used 'e' for the absolute term AN ENEMY and *e* for the
relative ENEMY OF, so it is quite difficult to confuse these two terms. It is Lewis who
confuses relative terms for absolute terms in the example quoted previously when he
takes *l* the relative term WHATEVER IS A LOVER OF _____ for the absolute term A LOVER.

We have above noted how Peirce regarded any absolute term as a relative term by
using a comma.

He held that such a relative term is a "relative formed by a comma" and defined
multiplication between absolute terms as multiplication between relatives using an
absolute term followed by an infinite series of commas. He wrote about this on page
374 of *Logic of Relatives*.

And if we are to suppose that absolute terms are multipliers at all (as
mathematical generality demands that we should), we must regard

every term as being a relative requiring an infinite number of correlates to its virtual infinite series "that is _____and is _____and is
_____etc." ...

Any term may be regarded as having an infinite number of factors, those at the end being ones, thus,

$$l, sw = l, sw, 1,1,1,1,1,1,1, \ldots$$

Peirce is justified in blurring the distinction between absolute terms and relative
terms, as absolute terms can be regarded as relatives by use of the comma multiplication. However it is usually clear from the context in *Logic of Relatives* whether the
absolute term or the relative version of it is intended. Brink notes:[27]

> Peirce feels free to use an absolute term as the multiplicand in a
> relative product: if R is a relation and X a class then
>
> $$R; X = \{x | (\exists y) [\langle x, y \rangle \in R]\}$$
>
> Coupled with the distinction between absolute and relative
> terms this leads to the conclusion that Peirce's is a multisystem: its
> operations can combine terms belonging to different categories.

This set-theoretical interpretation is of course anachronistic, but it should be noted
that not only can Peirce's operations combine terms but also they can actually trans-
form them from one category to another. In order to have the absolute term as the
multiplier then the absolute term must be converted to a relative term by logical multiplication or as Peirce wrote on page 372 of *Logic of Relatives*:

> Since our conception of multiplication is the application of a rela-
> tion, we can only multiply absolute terms by considering them as
> relatives.

Relative multiplication or the application of the relative term is then used to multiply such relative terms together and so the logical product according to Peirce, is a
special case of the relative product. He wished to connect the relative product, which is
of course, non-commutative with the logical or comma product which is commutative,
in order to include arithmetical algebra within his algebraic logic. It is evident that the

---

[27]See [**18**, p. 288].

demands of mathematical generality are very important to Peirce. He also emphasized the fact that this is not a different multiplication but a modification of it, i.e. logical multiplication is only functional multiplication with the addition of an extra correlate. So logical multiplication by an absolute term is in fact relative multiplication by the transformed relative term, and similarly logical multiplication by a relative term is relative multiplication by a conjugative term. He writes in a footnote to page 375:

> [logical multiplication] is multiplication by a relative whose meaning – or rather whose syntax – has been slightly altered; and that the comma is really the sign of this modification of the foregoing term.

However Brink has noted that it is incorrect to identify the relative product or the application of a relation, with logical multiplication as relative or functional multiplication is not commutative whereas logical multiplication is commutative.[28]

Burch takes a different approach in dealing with Peirce's method of application of a relation by considering absolute terms as relatives with only a first correlate. For example, the relation signified by LOVER OF _____ is indicated by the well formed formula x|Lxy in first order predicate logic where x is the first correlate and y the second, and the relative symbolized by HORSE is z|Hz so that the relative signified by LOVER OF (A) HORSE is

$$x \mid (\exists t)(Lxt \ \& \ Ht)$$

This is a binary (or dyadic) operation (APP2 as defined by Burch), in which two variables are identified and then quantified over existentially in a conjunction. It is symbolized graphically in Figure 4.1.

FIGURE 4.1.   LH

However the problem with symbolizing absolute terms as relative terms in this way is shown in the example of the conjugative term $\mathfrak{g}$ when used with the relative o and the absolute term h, this means as we have previously seen, GIVER OF A HORSE TO AN OWNER OF THAT HORSE. Here h is used in the sense of an individual horse rather than a class, and $\mathfrak{g}oh$ cannot be expressed by the symbolic means as employed above

---

[28]See [**18**, p. 290].

by Burch to illustrate the application of a relation. Instead a new operation must be devised of which the graphical representation is given by Burch in Figure 4.2.

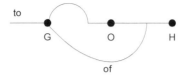

FIGURE 4.2.   GOH

4.3.2. Errors in *Logic of Relatives*. On page 400 Peirce supplies an "omission in the account given above of involution in this algebra." This section was added just before the paper went to the printers, when Peirce states that he saw it was necessary from his reading of De Morgan's "On the Syllogism: IV." It is therefore not surprising that this part of the work was not extensively revised by Peirce, with many proofs containing attributions to the wrong formulae.

Peirce introduces *backward involution*, along the lines of the converse of ordinary involution. For ordinary involution, l^s denotes WHATEVER IS A LOVER OF EVERY SERVANT. Note that Pierce is using servant in the sense of the relative term "servant of" rather than the absolute term "servant". So Peirce writes on page 401 EVERY LOVER OF WHATEVER SERVANT THERE IS where $^l s$ is interpreted as whatever is A LOVER OF NONE BUT SERVANTS.

(§124) $l^s = 1 - (1 - l)s.$

This follows from $(1 - x)y = 1 - xy$ where x is infinitesimal, used in the proof of §112 with $x = 1 - l$ and $y = s$, rather than §112 cited by Peirce which is

$$x = x - \tfrac{1}{2!}x^2 + \tfrac{1}{3!}x^3 - \tfrac{1}{4!}x^4 \ldots$$

Similarly, §125

$$^l s = 1 - l(1 - s).$$

The fundamental formulæ of backward involution are

(§126) $l_{(sw)} = (ls)_w$

(§127) $l +_, s_w = l_w, s_w$

(§128) $l_{(f,u)} = l_{f,}l_{u,}$

or as Peirce writes on page 402:

THE THINGS WHICH ARE LOVERS TO NOTHING BUT FRENCH VIOLINISTS ARE
THE THINGS THAT ARE LOVERS TO NOTHING BUT FRENCHMEN AND LOVERS TO
NOTHING BUT VIOLINISTS

(§128) $$l_{(f,u)} = \sigma^{-l(1-f,u)} = \sigma^{-(l(1-f) \,+,\, l(1-u))}$$

is proved as follows:

Peirce had previously defined σ^{-x} as $1 - x$. Here $(1 - f) +, (1 - u) = 1 - f, u$ by §125 as above and §29 which states,

$$(x - y) +, (z - w) = (x +, z) - (y +, w) + y, z, (1 - w) + x, (1 - y), w.$$

Putting $x = 1, y = f, z = 1$ and $w = u$ in the latter equation and using §22 $x +, x = x$ to give $f, u +, f, u = f, u$. Peirce is incorrect to cite §30 here rather than §29 since §30 as discussed in Section 4.1 is an equation based on Boole's development theorem.

We also have

$$
\begin{aligned}
l_f, l_u &= \sigma^{-l(1-f)}, \sigma^{-l(1-u)} \ldots &\quad §125 \\
&= \sigma^{-l(1-f) \,+,\, -l(1-u)} \ldots &\quad §11 \\
&= \sigma^{-(l(1-f) \,+,\, l(1-u))} \ldots &\quad §7
\end{aligned}
$$

So that

$$l_{(f,u)} = l_f, l_u.$$

Peirce is in error here when he cites §125, §13 and §7, since §11 should be used instead of §13 which is $(x, y)^z = x^z, y^z$.

In the following proof, given on page 403, of the theorem

(§131) $$l(u, f) = (lu), \Sigma_p((l - p)u +, pf), (lf).$$

I shall now analyze in some detail the steps and past theorems that Peirce uses to arrive at this conclusion and also supply omitted inferences and equations.

He first cites the equation $x_y = (1 - x)(1 - y)$ as derivable from §124 and §125. Peirce is again using $\sigma^{-xy} = (1 - x)^y$ from the proof of §112 which is however, not cited.

From §125,

(i)
$$x_y = 1 - x(1 - y)$$
$$= \sigma^{x(1-y)} = (1 - x)^{(1-y)}$$

by the proof of §112. We also have

(ii)
$$1 - (u +\!\!, f) = \sigma^{-(u +\!\!, f)} = \sigma^{-u}, \sigma^{-f} = (1 - u), (1 - f)$$

This follows from §11 $x^{y +\!\!, z} = x^y, x^z$ and §111 $\sigma^{-x} = 1 - x$. Peirce omits these details. So

$$(1 - l)^{(1-u)(1-f)} = (1 - l)^{1-(u +\!\!, f)}$$

and from (i)

$$(1 - l)^{(1-u)(1-f)} = {}^l(u +\!\!, f).$$

From the binomial theorem for backwards involution §129, rather than §128 $^l(f, u) = l_{f}.l_u$ as quoted by Peirce, we have

$$(1 - l)^{(1-u)(1-f)} = l_u + \Sigma_p{}^{lp}u, {}^p f +^l f$$

From (i)

$$(1 - l)^{(1-u)(1-f)} = (1 - l)^{(1-u)} + \Sigma_p(1 - (l - p))^{(1-u)}, (1 - p)^{(1-f)} + (1 - l)^{(1-f)}$$

Substituting u for $(1 - u)$ and f for $(1 - f)$ we get,

$$(1 - l)^{u,f} = (1 - l)^u +\!\!, \Sigma_p(1 - (l - p))^u, (1 - p)^f +\!\!, (1 - l)^f$$

By §124 $l^s = 1 - (1 - l)s$, so $(1 - l)^{u,f} = 1 - l(u, f)$ and also by §124 the left-hand side becomes:

$$1 - lu +\!\!, \Sigma_p(1 - (l - p)u), (1 - pf) +\!\!, 1 - lf.$$

Taking the contrary, which Peirce refers to as *taking the negative*, and using equation (ii) $1 - (u +\!\!, f) = (1-u), (1-f)$ which shows that the complement of a sum is the product of the complements, and by putting $1 - u = u$ and $1 - f = f$ that the complement of a product is the sum of the complements, we obtain:

$$l(u, f) = (lu), \Pi'_p((l - p)u +\!\!, pf), (lf).$$

Peirce interprets §131 as

THE LOVERS OF FRENCH VIOLINISTS ARE THOSE PERSONS WHO, IN REFER-
ENCE TO EVERY MODE OF LOVING WHATEVER, EITHER IN THAT WAY LOVE
SOME VIOLINISTS OR IN SOME OTHER WAY LOVE SOME FRENCHMEN.

The proof of §132 $(e,c)f = \Pi'_p(e(f-p) +\!\!, cp)$ is more straightforward since the step $^x y = (1-x)^{(1-y)}$ is not required to convert backward involution into ordinary involution and an extra replacement of complements as in (iii) is also avoided. Note that the terms ef and cf have been omitted although the formula when developed as in the previous proof to §131 can read

(§132) $(e,c)f = ef, \Pi'_p(e(f-p) +\!\!, cp), cf.$

In a later subsection entitled "Not" on page 420, Peirce looks back to the traditional syllogistic examples particularly those used by Boole, and re-expresses them using his notation of relative terms. He begins by supplying the missing formulae needed for the principles of contradiction and excluded middle. Using σ^{-x} rather than $(1-x)$, these are

$$x, \sigma^{-x} = 0;$$
$$x + \sigma^{-x} = 1.$$

A deduced property is

(§168) $\sigma^{-x^y} = \sigma^{-x} y.$

Peirce provides on page 421 the relevant interpretation INDIVIDUALS NOT SERVANTS OF ALL WOMEN ARE THE SAME AS NON-SERVANTS OF SOME WOMEN. It is interesting to note that Peirce uses the phrase SOME WOMEN here rather than SOME WOMAN as he did on page 378 or to explain §107 $l^s w \prec l^{sw}$. This is an example of the easy variation between a relative term y as in the formula (168) and the absolute term w used in the interpretation that has led to confusion as to the exact intentions of Peirce. However, although clear about the distinction between absolute terms and relative terms, Peirce by means of his logical product ',' the comma operation, can use such terms interchangeably.

Let us now consider errors in one of the two examples of syllogistic reasoning that Peirce gives in *Logic of Relatives*. Continuing his reflections on Boole's logic which was begun in "Harvard Lecture VI," he commented on page 422 that Boole cannot well describe hypothetical (EITHER ... OR, IF ... THEN) propositions or particular (using the quantifier SOME) propositions. He then proceeded to outline his first illustration of syllogistic reasoning taken from a Boolean example:

v,y=v(1-x)   SOME Ys ARE NOT Xs

then

$$v,y=v-v,x.$$

We can deduce

$$v,x=v=v,y-v,(1-y) \qquad \text{Some Xs are not Ys}$$

It is pointed out that this is not a valid deduction. So in the same example, Peirce showed firstly that Boolean syllogistic reasoning can be extended to relative terms and also that Boole could not accurately express such propositions. He maintained that to express Boole's algebra in relative terms, an existential condition, i.e. a relative denoting CASE OF THE EXISTENCE OF _____ or WHAT EXISTS ONLY IF THERE IS NOT _____ is required.

For example, in the equations $A = 0$ and $B = 0$, A and B represent respectively the propositions IT LIGHTENS and IT THUNDERS which in Boolean logic can take the values either 0 or 1, but are both equal to 0 in this case. For any x such that φx vanishes when x does not and vice versa, $\varphi A \prec \varphi B$ expresses the fact that IF IT LIGHTENS, IT THUNDERS. 0^x is such a function, 0 being the zero relative term such that $x + 0 = x$ and $x, 0 = 0$, vanishing when x does not, and not vanishing when x does. Zero, therefore can be interpreted as denoting THAT WHICH EXISTS IF, AND ONLY IF, THERE IS NOT _____ . So $0^x = 0$ means that there is nothing which exists if, and only if, some x does not exist.

Similarly for equations $A = 1$ and $B = 1$ meaning IT LIGHTENS and IT THUNDERS, where A and B have the value 1, we can use 0^{1-x} or ^1x as the function of x which vanishes unless x is 1 and does not vanish if x is 1.

Peirce comments on page 424:

We must therefore interpret 1 as THAT WHICH EXISTS IF, AND ONLY IF, THERE IS _____ ^1x AS THAT WHICH EXISTS IF, AND ONLY IF, THERE IS NOTHING BUT x, and 1x AS THAT WHICH EXISTS IF, AND ONLY IF, THERE IS SOME x.

Then if the propositions X and Y are $A = 1$ and $B = 1$, the four equivalent forms of hypothetical propositions can be expressed in terms of the logic of relatives shown in Table 4.1.

TABLE 4.1. Hypothetical Propositions

IF X, THEN Y	$1A \prec 1B$
IF NOT Y, THEN NOT X	$1(1 - B) \prec 1(1 - A)$
EITHER NOT X OR Y	$1(1 - A) +, 1B - 1$
NOT BOTH X AND NOT Y	$1A, 1(1 - B) = 0$

Particular propositions involving the concept of SOME, are expressed by the contradictions of universal propositions, e.g., as $h, (1 - b) = 0$ means that EVERY HORSE IS BLACK, so $0^h(1 - b) = 0$ means that SOME HORSE IS NOT BLACK, and $h, b = 0$ means that No HORSE IS BLACK, so $0^{h,b} = 0$ means that SOME HORSE IS BLACK.

A second example of syllogistic logic using the logic of relations is then given on page 425: Given the premises, EVERY HORSE IS BLACK, and EVERY HORSE IS AN ANIMAL, we require the conclusion. We are given:

$$h \prec b; \qquad h \prec a.$$

Commutatively multiplying we get, $h \prec a, b$. Then by §92 If $a \prec b$ then $c^b \prec c^a$, we get $0^{a-b} \prec 0^h$, and by §40 $0^x = 0$, provided $x > 0$, i.e. for x non-zero, we have the conclusion

If $h > 0$ then $0^{a,b} = 0$, or IF THERE ARE ANY HORSES, SOME ANIMALS ARE BLACK.

Peirce wrote on page 425, "I think it would be difficult to reach this conclusion, by Boole's method unmodified." He also provided an alternative expression of particular propositions by using inequalities, e.g. writing SOME ANIMALS ARE HORSES as $a, h > 0$. The above problem, i.e. to find the result of the two premises EVERY HORSE IS BLACK and EVERY HORSE IS AN ANIMAL can be expressed by using the strict inequality $>$ rather than illation $\prec$. Peirce does not give a satisfactory account of this however and only briefly stated:

the conclusion required in the above problem might have been obtained in this form, very easily, from the product of the premises, by §1 and §21.

From this, I conjecture that he is supplying the reasoning for the multiplication process: If h ≺ b and h ≺ a then h ≺ a, b which follows from the rule of transitivity §1 If x ≺ y and y ≺ z, then x ≺ z, and also from §90 If a ≺ b, then ca ≺ cb and §91 If a ≺ b, then ac ≺ bc rather than §21 x ≺ x +, y which is a formula concerning commutative addition rather than commutative multiplication. §90 and §91 are amongst a group of formulae that are said to be derived from §1, §21 and §2 If x = y then φx = φy, although in this case §90 and §91 are derived from §1 and §27, which explains why §21 is mentioned although it is not in fact relevant.

The reasoning for if h ≺ b and h ≺ a then h ≺ a, b is as follows: h ≺ a so by §91 h, b≺a, b since absolute terms h, a and b can be regarded as relative terms by the comma multiplication, e.g., h, Horse that is _____. But since h ≺ b so by §90 h, h ≺ h, b or h ≺ h, b. So by §1 we have h ≺ h, b and h, b ≺ a, b therefore h ≺ a, b. Since h ≺ a, b it follows that if h > 0 then a, b > 0 or If there are any horses then some animals are black.

4.4. Peirce's "Differentiation"

The process of differentiation introduced by Peirce in *Logic of Relatives*, has for many years proved fairly obscure. D. Merrill writes in his introduction to [56]:

> ... the subsection on Infinitesimal Relatives contains the most elaborate mathematical analogies in the memoir, with very puzzling applications of such mathematical techniques as functional differentiation and the summation of series.

In order to understand properly Peirce's differentiation technique we must first consider

a) the meaning of numerical coefficients,
b) the binomial theorem as developed by Peirce,
c) infinitesimal relatives,

d) the number associated with a logical term.

4.4.1. Numerical Coefficients, the Binomial Theorem and Infinitesimal Relatives in *Logic of Relatives*.

The numerical coefficient of a relative term is introduced through the concept of the subscript number or *subjacent* number associated with a conjugative term that indicates how far to the right is the first correlate. Since logical multiplication forms either relative terms from absolute terms and conjugative terms from relative terms, Peirce introduces two subjacent numbers 0 and ∞ to be used with this multiplication. These serve to indicate the first correlate and we have $s,_0 w = sw$ so that 0 neutralizes a comma, and by removing the correlate to infinity so as to leave it indeterminate, we have m, ∞ as an expression for SOME MAN and l_∞ or 1 to express SOMETHING and l_0 or 1 to express ANYTHING. The parallel is drawn with Boole's unity or WHATEVER IS. This multiplication is commutative as in $s, l = l, s$. Using the infinity subscript notation for SOME, Peirce arrives at $x,_\infty + x,_\infty = 2.x,_\infty$ or more simply $2x$, (although Peirce more often uses $2.x$), meaning SOME x together with SOME x equals SOME TWO xs where the '.' signifies invertible multiplication, i.e. that the xs are disjoint. This type of multiplication is not commutative, e.g., $l2.w$ meaning THE LOVERS OF SOME TWO WOMEN and $2.lw$ meaning SOME TWO LOVERS OF A WOMAN are not the same. 2 is used to denote two individual things.

Consider also how Peirce develops the binomial theorem by means of relative terms. For the non-disjoint relative terms emperor and conqueror, we have for the class EMPEROR OR CONQUEROR OF ALL FRENCHMEN, that this is equal to THE EMPEROR OF ALL FRENCHMAN, or EMPEROR OF SOME FRENCHMEN AND CONQUEROR OF THE REST, or CONQUEROR OF ALL FRENCHMEN, so that

$$(e +, c)^f = e^f +, \Sigma_p e^{f-p}, c^p +, c^f.$$

To explain the term Σ_p he writes:

> Σ_p denotes that p is to have every value less than z, and is to be taken out of z in all possible ways, and that the sum of all the terms so obtained of the form e^{f-p}, c is to be taken.[29]

---

[29]See [**149**, p. 362].

An alternative equation is also given:

$$(e \,+\!, c)^f = e^f \,+\!, [f] \cdot e^{f-\dagger 1}, c^{1\dagger} \,+\!, \frac{[f] \cdot ([f] - 1) \cdot e^{f-\ddagger 2}, c^{2\ddagger} + \ldots}{2}$$

where $[f]$ stands for the number of individuals represented by the absolute term f, a representative FRENCHMAN, and $e^{f-\dagger 1}, c^{1\dagger}$ stands for the class of EVERYTHING WHICH IS AN EMPEROR OF EVERY FRENCHMAN BUT SOME ONE FRENCHMAN AND IS AN EMPEROR OF THAT ONE. This works equally well for disjoint addition and either term of the binomial may be negative provided that we assume $(-x)^y = (-)^{[y]}.x^y$. For example, from §129 we have

$$^l(u \,+\!, f) = {}^l u + \{l\}^{l-\dagger 1_u, 1\dagger_f} + \{l\}.(\{l\} - 1\}/2)^{l-\ddagger 2_u, \ddagger 2_f} + \ldots$$

where $\{l\}$ represents the NUMBER OF PERSONS THAT ONE PERSON IS LOVER OF, rather than $[l]$ which is the AVERAGE NUMBER OF LOVERS OF ONE INDIVIDUAL, as used in ordinary involution. The literal translation provided by Peirce is

> THOSE PERSONS WHO ARE LOVERS OF NOTHING BUT FRENCHMEN AND VIO-
> LINISTS CONSIST FIRST OF THOSE WHO ARE LOVERS OF NOTHING BUT FRENCH-
> MEN; SECOND, OF THOSE WHO IN SOME WAYS ARE LOVERS OF NOTHING BUT
> FRENCHMEN AND IN ALL OTHER WAYS OF NOTHING BUT VIOLINISTS, AND FI-
> NALLY OF THOSE WHO ARE LOVERS ONLY OF VIOLINISTS.

Finally infinitesimal relatives[30] are introduced on page 395. If x is taken to have only one individual correlate, then $x^2 = 0$, i.e. there is no relative x that stands in this relation to two individuals, since x has only one correlate which is an individual. This is an existence property and x is called *infinitesimal* because of the vanishing of its powers. Consider x to have only two individual correlates then again $x^3 = 0$ and all its higher powers vanish, since x does not stand in that relation to more than two individuals. Here x is again an infinitesimal relative. Peirce writes on page 391, "those relatives whose correlatives are individual: I term these infinitesimal relatives." Obviously the analogy is to the infinitesimals δx of Leibniz in the differential calculus. As we shall see we need the relative Δx, (although not x), to be an infinitesimal relative in Peirce's differentiation process.

---

[30]Martin has noted the analogy between infinitesimal relatives and many-one relations. See [**121**, pp. 34–38].

For x infinitesimal, we have the binomial theorem $(1 + x)^n = 1 + xn$. Putting i for x because x is an infinitesimal relative we have:

$$xn = in = y,$$

we have

$$(1 + i)^{y/i} = 1 + y$$

If $y = 1$ then Peirce defines $\sigma = (1 + i)^{1/i} = 1 + 1$.

Positive powers of σ are "absurdities," according to Peirce, i.e. they are trivial. This is because $\sigma^x = 1 + x = 1$. But for negative powers we have, §111 $\sigma^{-x} = 1 - x$. This has the meaning WHATEVER IS OTHER THAN EVERY x; so that σ^- means NOT. Peirce defined $\log x$ by the equation

$$\sigma^{\log x} = x$$

By §111 and §10 $(x^y)^z = x^{(yz)}$,

$$\sigma^{-xy} = (1 - x)^y = 1 - xy.$$

Looking at the binomial development of $(1 - a)^x$

$$(1 - a)^x = 1 - [x] \cdot 1^{x - \dagger 1}, a^{\dagger 1} +, (([x] \cdot [x - 1])/2).1^{x - \ddagger 2} a^{\ddagger 2} +, \cdots .$$

Peirce then used the notation $(ax)^3$ for

$$([x] \cdot [x - 1] \cdot [x - 2]) \cdot 1^{x - \dagger 3}, a^{\dagger 3}$$

"that is for whatever is a to any three x's, regard being had for the order of the x's," and using the numbers as exponents.

Applying the binomial theorem,

$$(1 - a)^x = 1 - ax + \frac{1}{2!}(ax)^2 - \frac{1}{3!}(ax)^3 + \cdots$$

$$= 1 - a\left(x - \frac{1}{2!}x^2 + \frac{1}{3!}x^3 - \frac{1}{4!}x^4 \cdots\right).$$

But since x is infinitesimal we have the higher powers of x vanishing so that

$$(1 - a)^x = 1 - ax.$$

So §112

$$x = x - \frac{1}{2!}x^2 + \frac{1}{3!}x^3 - \frac{1}{4!}x^4 \cdots$$

Peirce is emphasizing here the analogy with the power series for e^{-x}.

4.4.2. The Algebraic Development of Peirce's Differentiation. In this next section Peirce introduces his mathematically analogous process of differentiation by introducing Δ, a difference operation on relative terms. He establishes the algebraic basis for the differentiation using his notation of relative and infinitesimal terms. I have included more detailed explanation where such details have been omitted and also noted small errors, which show that Peirce did not extensively revise this section. The original text is enclosed in quotation marks to distinguish it from my additional comments and proofs.

The first difference of a function is defined by the usual formula:

(§113) $$\Delta\varphi x = \varphi(x + \Delta x) - \varphi x$$

where we also have §114 $x, (\Delta x) = 0$ and $x + \Delta x = x +, \Delta x$. However it has been previously stated by Peirce that φx is the Boolean function in which φ is a function in x involving only the commutative operations and the operations inverse to them, e.g. for addition we could have $\varphi x = x + x$. §114 ensures that Δx defined as an indefinite relative never has a correlate in common with the relative term x. Higher differences are then defined by the formulæ

(§115) $$\Delta^n \cdot x = 0 \text{ if } n > 1$$

But $\Delta^{n\cdot}$ means apply Δ n times. As Peirce stated on page 398:

> The exponents here affixed to Δ denote the number of times this operation is to be repeated, and thus have quite a different signification from that of the numerical coefficients in the binomial theorem. I have indicated the difference by putting a period after exponents significative of operational repetition. Thus, m^2 may denote a MOTHER OF A CERTAIN PAIR and $m^{2\cdot}$ a MATERNAL GRANDMOTHER.

It should be noted that §115 also shows that Δx is an infinitesimal relative.[31]

$$\Delta^{2\cdot}\varphi x = \Delta\Delta\varphi x = \varphi(x + 2.\Delta x) - 2.(\varphi(x + \Delta x) + \varphi x$$

---

[31]Peirce has incorrectly $\Delta\Delta x$ for $\Delta\Delta\varphi x$ and similarly $\Delta\Delta^{2\cdot} x$ for $\Delta\Delta^{2\cdot}\varphi x$.

This follows because

$$\Delta\Delta\varphi x = \Delta(\varphi(x + \Delta x) - \varphi x)$$
$$= \varphi(x + \Delta x + \Delta x) - \varphi(x + \Delta x) - (\varphi(x + \Delta x) - \varphi x)$$
$$= \varphi(x + 2.\Delta x) - 2.\varphi(x + \Delta x) + \varphi x$$

Similarly,

$$\Delta^{3}\varphi x = \Delta\Delta^{2}\varphi x = \varphi(x + 3.\Delta x) - 3.\varphi(x + 2.\Delta x) + 3.\varphi(x + \Delta x) - \varphi x$$

In general §116,

$$\Delta^{n}\varphi x = \varphi(x + n.\Delta x)$$
$$- n.\varphi(x + (n - 1).\Delta x)$$
$$+ n.((n - 1)/2).\varphi(x + (n - 2).\Delta x)\ldots$$

Peirce then defined the limiting process on page 398:

> If Δx is relative to so small a number of individuals that if the number were diminished by one, $\Delta^{n}\varphi x$ would vanish, then I term these two corresponding differences differentials, and write them with d instead of Δ.
>
> This ensures that the least number of correlates is taken to ensure existence of the differentiation process. The difference of the invertible sum of two functions is then shown to be the sum of their differences.

(§117)
$$\Delta(\varphi x + \psi x) = \varphi(x + \Delta x) + \psi(x + \Delta x) - \varphi x - \psi x$$
$$= \varphi(x + \Delta x) - \varphi x + \psi(x + \Delta x) - \psi x$$
$$= \Delta\varphi x + \Delta\psi x$$

This follows from §113 and §17 rather than §113 and §18 as cited by Peirce.

In the next section he produced four equations labeled under §118. If a is a constant [relative] then,

(§118)
$$\Delta a\varphi x = a(\varphi x \,+\!, \Delta\varphi x) - a\varphi x$$
$$= a\Delta\varphi x - (a\Delta\varphi x), a\Delta\varphi x$$
$$\Delta^{2\cdot} a\varphi x = -\Delta a\varphi x, a\Delta x, \ldots$$
$$\Delta(\varphi xa) = (\Delta\varphi x)a - ((\Delta\varphi x)a), \varphi xa, {}^{32}$$
$$\Delta^{2\cdot}(\varphi xa) = -(\Delta\varphi x)a, \ldots {}^{33}$$

However the first equation should read
$$\Delta a\varphi x = a(\Delta\varphi x + \varphi x) - a\varphi x = a\Delta\varphi x.$$

We have
$$\Delta a\varphi x = a\varphi(x + \Delta x) - a\varphi x$$
from the definition of $\Delta\varphi x$ and since
$$\Delta\varphi x = \varphi(x + \Delta x) - \varphi x$$
then it follows that
$$\varphi(x + \Delta x) = \Delta\varphi x + \varphi x$$
and so
$$\Delta a\varphi x = a(\Delta\varphi x + \varphi x) - a\varphi x$$

This gives $\Delta a\varphi x = a\Delta\varphi x$ from §5 $x(y + z) = xy + xz$. Note that this differs from Peirce's result in §118 as he uses the non-invertible addition ' $+\!,$ ' but this is incorrect if the definition of $\Delta\varphi x$ specifies invertible operations only. By writing $a\varphi x \,+\!, a\Delta\varphi x$ for $a\varphi x + a\Delta\varphi x$ Peirce has omitted the extra term of $+(a\Delta\varphi x), a\varphi x$ (this follows from §24 $x \,+\!, y = x + y - xy$). So that the correct reading should be $\Delta a\varphi x = a\Delta\varphi x$ instead of the result obtained in *Logic of Relatives* of $\Delta a\varphi x = a\Delta\varphi x - (a\Delta\varphi x), a\varphi x$.

Peirce then has
$$\Delta^{2\cdot} a\varphi x = -\Delta a\varphi x, a\Delta x, \ldots$$
However
$$\Delta^{2\cdot} a\varphi x = \Delta\Delta a\varphi x = \Delta(a\varphi(x + \Delta x) = a\varphi x) = \Delta a\varphi(x + \Delta x) - \Delta a\varphi x$$

from §117. But using our result of $\Delta a\varphi x = a\Delta\varphi x$

$$\Delta a\varphi(x + \Delta x) - \Delta a\varphi x = a\Delta\varphi(x + \Delta x) - a\Delta\varphi x$$
$$= a(\Delta\varphi(x + \Delta x) - \Delta\varphi x)$$
$$= a\Delta(\varphi(x + \Delta x) - \varphi x)$$
$$= a\Delta\Delta\varphi x$$
$$= a\Delta^{2\cdot}\varphi x$$

so that we obtain

$$\Delta^{2\cdot} a\varphi x = a\Delta^{2\cdot}\varphi x$$

It is possible that Peirce wrote $-\Delta a\varphi x, a\Delta x \dots$ for $-\Delta a\varphi x, a\varphi x \dots$ He does not repeat the $a\Delta x$ term in the last equation §119,

$$\Delta(a, \varphi x) = a, \Delta\varphi x.$$

The reasoning being

$$\Delta(a, \varphi x) = a, \varphi(x + \Delta x) - a, \varphi x$$
$$= a, (\varphi(x + \Delta x) - \varphi x)$$
$$= a, \Delta\varphi x$$

from the dual formula to §19 $x, (y + z) = xy + x, z$, given by Peirce, but with invertible subtraction instead of addition.

The differentiation process is developed by example as follows:

$$\Delta(x^2) = (x + \Delta x)^2 - x^2 = 2.x^{2-\dagger 1}, (\Delta x)^{\dagger 1} + (\Delta x)^2$$

from the binomial theorem. Similarly

$$\Delta(x^3) = (x + \Delta x)^3 - x^3 = 3.x^{3-\dagger 1}, (\Delta x)^{\dagger 1} + 3.x^{3-\ddagger 2}, (\Delta x)^{\ddagger 2} + (\Delta x)^3$$

If Δx is infinitesimal, i.e. relative to only one individual then $(\Delta x)^2$ vanishes and we have writing d for Δ

$$d(x^2) = 2.x^1, d x$$

Similarly

$$d(x^3) = 3.x^2, d x$$

For the second differential by §115

$$\Delta^{2\cdot}(x^3) = (x + 2.\Delta x)^3 - 2.(x + \Delta x)^3 + x^3$$

On expansion by the binomial theorem,

$$\Delta^{2\cdot}(x^3) = 6.x^{3-\ddagger 2}, (\Delta x)^{\ddagger 2} + 6.(\Delta x)^3$$

If Δx is relative to less than two individuals, then $\Delta^{2\cdot}(\varphi x)$ vanishes[34] where $\varphi x = x^3$, so making it relative to two only we have,

$$d^2 \cdot (x^3) = 6.x^1, (dx)^2$$

From these examples we can see, where n is a logical term, then

$$\Delta(x^n) = (x + \Delta x)^n - x^n = [n].x^{n-\dagger 1}, (\Delta x)^{\dagger 1} + \dots$$

by the binomial theorem.

$$d(x^n) = [n].x^{n-\dagger 1}, (dx)$$

And we have

(§120) $d^{m\cdot}(x^n) = [n].[n-1].[n-2]\dots[n-m+1].x^{n-\dagger m}, (dx)^{\dagger m}$

Differentiating l^x

$$\Delta l^x = l^{x + \Delta x} - l^x, = l^x, l^{\Delta x} - l^x = l^x, (l^{\Delta x} - 1)$$

since by §11

$$x^{y + z} = x^y, x^z$$

But by §111, $\sigma^{-x} = 1 - x$ so

$$\sigma^{l^{\Delta x} - 1} = l^{\Delta x}$$

and hence

$$l^{\Delta x} - 1 = \log l^{\Delta x}$$

It can be shown that

$$\log l^{\Delta x} = (\log l)\Delta x$$

the proof follows from the usual law of logarithms and using §10 $(x^y)^z = x^{(yz)}$, so we have §121,

$$d l^x = l^x, (\log l) d x^{35}$$
$$= l^x, (l - 1) d x$$
$$= -l^x, (1 - l) d x$$

since $\sigma^{l-1} = l$ implies $l - 1 = \log l$.

---

[34]Peirce has "$\Delta\varphi x$ vanishes" here.
[35]Peirce has $\log, l$ for $\log l$.

After setting up this complex notation for his version of logical differentiation Peirce sought to apply the process to two specific areas connected with the calculus namely Maclaurin's theorem and maxima and minima problems. On page 406 of *Logic of Relatives*, Peirce writes Maclaurin's theorem in the following way:

(i) $$x = \frac{x}{dx}\frac{0}{x}\left(\frac{1}{0!}d^{0.} + \frac{1}{1!}d^{1.} + \frac{1}{2!}d^{2.} + \frac{1}{3!}d^{3.} + \ldots\right)\varphi x$$

where $\frac{x}{dx}$ means replace x by dx in the formula, $\frac{0}{x}$ means replace 0 with x. Apart from the fact that Peirce has incorrectly written $x =$ at the beginning of this equation for $\varphi x =$, the reason the equation holds is that Maclaurin's theorem can be written:

(ii) $$\varphi x = \varphi 0 + x.\varphi' 0 + x^2.\varphi''\frac{0}{2!} + x^3.\varphi'''\frac{0}{3!} \ldots$$

and we can obtain

(iii) $$d^{0.}\varphi x = \varphi x$$

$$d^{1.}\varphi x = \varphi' x, dx$$

$$d^{2.}\varphi x = \varphi'' x, (dx)^2$$

$$\ldots$$

and from §120

$$d^{m.}(x^n) = [n].]n - 1].[n - 2]\ldots[n - m + 1].x^{n - \dagger m}, (dx)^{\dagger m}$$

Then substituting into Peirce's equation (i)

$$\varphi x = \frac{1}{0!}d^{0.}(\varphi x) + \frac{1}{1!}d^{1.}(\varphi x) + \frac{1}{2!}d^{2.}(\varphi x) + \frac{1}{3!}d^{3.}(\varphi x) + \cdots$$

so we obtain using (iii)

$$\varphi x = \frac{1}{0!}\varphi x + \frac{1}{1!}\varphi' x, dx + \frac{1}{2!}\varphi'' x, (dx)^2 + \frac{1}{3!}\varphi''' x, (dx)^3 + \cdots$$

Replacing on the right hand side x with 0 and dx with x gives the required result, i.e. Maclaurin's theorem as expressed in (ii).

Peirce also used differentiation in logic to solve a maximum and minimum problem, i.e. in a certain institution all the officers (x) and all their common friends (f^x) where Peirce writes f for f^x and f is the relative term 'FRIEND OF _____, are privileged persons (y). The question is to minimize y. Here Peirce applies differentiation not only to a relative (f) but also to classes x and y. However these classes can be thought of as relatives in the sense that x is replaced by x, or x THAT IS _____.

Peirce gave his definition of a minimum on page 406:

When y is at a minimum it is not diminished either by an increase or diminution of x.

He continued:

for $[dy] \prec 0$ when $[x]$, which is THE NUMBER OF OFFICERS, is diminished by one, $[dy] \prec 0$.

This has clear analogies with the minimum of the differential function in calculus.

Since
$$y = x + f^x, dy = dx + df^x = dx - f^x, (1 - f)dx,$$
from §121. So when y is a minimum[36]
$$[dx - f^x, (1 - f)dx] > 0 \quad [dxf^{x-1}, (1 - f)dx] \prec 0$$
So,
$$[dx] - [f^x, (1 - f)dx] \prec 0 \quad [dx] - [f^{x-1}, (1 - f)dx] \prec 0$$
Peirce now states that by §30
$$\varphi x = (\varphi 1), x + (\varphi 0)(, 1 - x),$$
the development theorem,
$$f^x, (1 - f)dx = f^x - (0; 0), (1 - f)dx$$
This is the final result. But it can be demonstrated that the last term $(0; 0), (1 - f)dx$ is incorrect. The reasoning behind this can now be traced. Consider
$$f^x = \frac{f^x, (1 - f)dx}{(1 - f)dx}$$
Here we have $a = b/c$, where $a = f^x$, $b = f^x, (1 - f)dx$, and $c = (1 - f)dx$. Using the development theorem §30 applied to two symbols $a = \varphi(b, c)$ we have,
$$a = b, c + 0/0, (1 - b), (1 - c) + 0/1, (1 - b), c + 1/0, b, (1 - c).[37]$$

---

[36]Peirce mistakenly writes here "when x is a minimum."
[37]Peirce has used this form of the development theorem before in "Harvard Lecture VI"[55, p. 233].

So,

$$f^x = f^x, (1 - f)dx, (1 - f)dx + 0/0, (1 - f^x, (1 - f)dx), (1 - (1 - f)dx)$$
$$+ 0/1, (1 - f^x, (1 - f)dx), (1 - f)dx + 1/0, f^x, (1 - f)dx, (1 - (1 - f)dx).$$

However the last two terms vanish since firstly $0/1 = 0$ and secondly[38]

$$(1 - f)dx, (1 - (1 - f)dx) = 0.$$

So,

$$f^x = f^x, (1 - f)dx, (1 - f)dx + 0/0, (1 - f^x, (1 - f)dx), (1 - (1 - f)dx)$$
$$= f^x, (1 - f)dx + 0/0, (1 - f^x, (1 - f)dx), (1 - (1 - f)dx)$$

as $x, x = x$ from §23. Now consider

$$(1 - f^x, (1 - f)dx), (1 - (1 - f)dx).$$

Since $f^x, (1 - f)dx \prec (1 - f)dx$, from §94 we have

$$(1 - f^x, (1 - f)dx), (1 - (1 - f)dx) = 1 - (1 - f)dx$$

This follows because $a, b \prec b$, and so $(1 - a, b), (1 - b) = 1 - b$. So

$$f^x = f^x, (1 - f)dx + 0/0, (1 - (1 - f)dx)$$

and therefore,

$$f^x, (1 - f)dx = f^x - 0/0, (1 - (1 - f)dx)$$

rather than Peirce's result.

$$f^x, (1 - f)dx = f^x - 0/0, (1 - (1 - f)dx)$$

also agrees with the definition for logical division $a/b = a + 0/0(1 - b)$ which Peirce gave in "Harvard Lecture VI,"[39] where

$$a = f^x, (1 - f)dx \text{ and } b = (1 - f)dx.$$

$$f^x = \frac{f^x, (1 - f)dx}{(1 - f)dx} = f^x, (1 - f)dx + 0/0(1 - (1 - f)dx)$$

thus

$$f^x, (1 - f)dx = f^x - 0/0(1 - (1 - f)dx)$$

---

[38]See [**149**, p. 421] for another instance of this written as $x, \sigma^{-x} = 0$.
[39]See 104 above.

4.4.3. A Logical Basis for Peirce's Theory of Differentiation. Having established the algebraic foundations of differentiation using relatives and infinitesimals, I now provide examples that demonstrate that there is a logical interpretation for this theory. Peirce's differentiation using relative terms is not only a convenient algebraic contrivance but also has a logical validation. Peirce however, concerned himself only with the algebraic equations, neglecting to give such interpretations. Consider the equation that we have seen above obtained by Peirce,

$$d(x^2) = 2.x^1, d\,x.$$

A distinction must be drawn between the exponents above obtained from the coefficients in the binomial theorem, which denote the number of individual correlates of the relative term x, and the exponents denoting the number of times an operation is to be repeated. Peirce states on page 398:

> I have indicated the difference by putting a period after exponents significative of operational repetition. Thus m^2 may denote a MOTHER OF A CERTAIN PAIR, $m^{2\cdot}$ a MATERNAL GRANDMOTHER.

To provide interpretations for the differentiation process let us take $\varphi x = x^2$. To find $d(\varphi x)$:

> Let x be the class consisting of WHATEVER IS A SERVANT OF _____.
> Let Δx be the class consisting of WHATEVER IS A LOVER OF TOM.
> Then x^2 will be the class consisting of WHATEVER IS THE SERVANT OF TWO INDI-
> VIDUALS, say x^2 is the class of SERVANTS OF JACK AND JILL.

We must ensure that x and Δx never have a correlate in common. This we can do by ruling out a SERVANT and LOVER of the same individual, i.e. a SERVANT OF JACK cannot be a LOVER OF JACK[40] so that $x, \Delta x = 0$. Δx is an infinitesimal relative. It has been defined as a LOVER OF TOM, i.e. a LOVER OF ONE PERSON ONLY, and since it has only one correlate $(\Delta x)^2 = 0$ and all higher powers vanish, where $(\Delta x)^2$ means the class of WHATEVER IS A LOVER OF A CERTAIN PAIR. An infinitesimal term x is a relative term such that higher powers vanish, i.e. no such x exists. Another necessary condition that should be borne in mind is that the number of correlates of Δx are required to be the least such number such that $\Delta x^{n\cdot}\varphi x$ exists.

---

[40]Morally unethical to the Victorians, but not perhaps to Peirce. See [**17**, p. 147].

From the previous section 4.4.2 we have seen,

$$\Delta(x^2) = (x + \Delta x)^2 - x^2 = 2.x^{2-\dagger 1}.(\Delta x)^{\dagger 1} + (\Delta x)^2$$

Since by limiting Δx to only one correlate, i.e. Tom, this means $(\Delta x)^2$ vanishes, so

$$\Delta(x^2) = 2.x^{2-\dagger 1}.(\Delta x)^{\dagger 1}$$

Obviously reducing the number of correlates of Δx by one would make $\Delta x(\varphi x)$ vanish so we can now replace Δx by $d x$ to obtain

$$d(x^2) = 2.x^1. d x$$

The interpretation is therefore that the differentiation process acting on the class of SERVANTS OF JACK AND JILL produces TWO SERVANTS OF JILL (ONLY) THAT ARE ALSO LOVERS OF JACK.

In a similar way, we can find $d(x^3)$, taking $\varphi x = x^3$. Let Δx be the class WHATEVER IS A LOVER OF TOM and x^3 the class of SERVANTS OF TOM, DICK AND HARRY. We cannot have one individual who is both a LOVER and SERVANT of the same individual. In the previous section it was shown that Peirce obtained algebraically from the binomial theorem,

$$\Delta(x^3) = 3.x^{3-\dagger 1}.(\Delta x)^{\dagger 1} + \left[3x^{3-\ddagger 2}\right].(\Delta x)^{\ddagger 2} + (\Delta x)^3$$

Limiting Δx to only one correlate then $(\Delta x)^{\ddagger 2}$ and $(\Delta x)^3$ vanish, so that

$$\Delta(x^3) = 3.x^{3-\dagger 1}.(\Delta x)^{\dagger 1}.$$

Since, if we restricted Δx to less one correlate then the entire expression for $\Delta x(x^3)$, i.e. $\Delta x(\varphi x)$ vanishes, so we can then replace Δx by $d x$ obtain,

$$d(x^3) = 3.x^2. d x$$

The interpretation is therefore that the differentiation process acts upon the class of SERVANTS OF TOM, DICK AND HARRY to produce a class of THREE SERVANTS OF DICK AND HARRY THAT ARE LOVERS OF TOM. Note that the differential coefficient obtained is the number of individual correlates specified in φx, i.e. 3.

The second differential is obtained as follows:

$$\Delta^2(x^3) = 6.x^{3-\ddagger 2}.(\Delta x)^{\ddagger 2} + 6.(\Delta x)^3$$

as in the previous section 4.4.3. Let $\varphi x = x^3$ as before. This time the number of correlates of Δx is restricted to two, so that Δx is the class of WHATEVER IS A LOVER OF

TOM AND DICK. Δx is an infinitesimal relative since it has individual correlates and so higher powers will vanish, i.e. not exist. With this restriction we now have,

$$\Delta^{2\cdot}(x^3) = 6.x^{3-\ddagger 2}, (\Delta x)^{\ddagger 2}$$

Since, if we further reduce the number of correlates of Δx by one, i.e. have Δx relative to only one individual, this means that $(\Delta x)^{\ddagger 2}$ and therefore the above expression for $\Delta^{2\cdot}(\varphi x)$ vanishes, so that this has fulfilled Peirce's condition – the number of correlates of Δx are required to be the least such number such that $\Delta x^{n\cdot}[\varphi x]$ exists, so that we can replace Δ by d to conclude

$$d^{2\cdot}(x^3) = 6.x^1, (d x)^2$$

The interpretation being that the second differential acts on the class of SERVANTS OF TOM, DICK AND HARRY to produce the class of SIX SERVANTS OF HARRY WHO ARE LOVERS OF TOM AND DICK. In summary,

$$d(x^n) = [n].x^{n-1}, d x$$

The differentiation process acting upon the class of xs of ns (where x and n are relative terms) results in a class of xs with a reduction in the number of correlates of each member by one individual, where each member is in the relation signified by the infinitesimal relative Δx to this one individual. The number of correlates of Δx is taken to be the least number that will give a non-zero result on differentiating. The differential coefficient obtained on differentiating x^n is the usual or standard number[41] of individuals associated with the class of ns.

4.5. Conclusions

Peirce began *Logic of Relatives* by setting out the case that the two most important logical theories of the time, De Morgan's logic of relations and Boole's algebraic theory of classes and propositions were inadequate. The stated object of this paper was to extend Boole's logic by incorporating De Morgan's relational logic to provide a complete theory of logic. He was however, not uncritical of De Morgan's work, describing it as a system which 'still leaves something to be desired'. Boole's algebraic logic is also criticized as being 'restricted to that simplest and least useful part of the subject, the logic of absolute terms'. Rather than develop the theory of relations in logic without reference to Boole, as De Morgan had done, Peirce wished to extend the Boolean system to include relations and in doing so, developed in this paper a new notation to

---

[41]Peirce used the word 'average' here, but in the non-mathematical sense of 'normal' or 'standard'.

deal with the task. Although he claimed only to introduce the notation, rather like his father Benjamin's claim to develop the 'language' rather than the 'grammar' of his linear algebras (hence the full title of the paper being "Description of a Notation for the Logic of Relatives") Peirce in fact does much more. He also developed formulae and ways of working with the notation, incorporating the binomial theorem, differentiation and much more within his theory. Peirce clearly intended that algebra would be able to be represented within his new notation. He wrote in *Logic of Relatives*, 'arithmetical algebra should be included under the notation employed as a special case of it'. Earlier he stated that he was guided by the analogy to algebra that inspired both Boole and De Morgan: 'As we are to employ the usual algebraic signs as far as possible, it is proper to begin by laying down definitions of the various relations and operations.'

The fact that Peirce sought to include so many mathematical theorems within his algebraic logic is a sign of his fundamental view that mathematics extends over the whole realm of formal logic. He was guided primarily by algebraic analogy. He wrote on page 360, 'In extending the use of old symbols to new subjects, we must of course be guided by certain principles of analogy, which, when formulated, become new and wider definitions of these symbols ... We ... employ the usual algebraic signs as far as possible'.

The analogy between mathematics and logic is a strong one. Peirce later listed on page 363 four conditions that relations and operations should possess:

1) It is an additional motive for using a mathematical sign if the general conception of the operation or relation resembles that of the mathematical one.
2) Numbers should be capable of being substituted for logical terms and the equations should still hold good, so that numerical algebra is included under the notation as a special case of it.
3) A zero and a unity term having similar mathematical properties, e.g., $x + 0 = x$ should be possible.
4) A strong motive for the adoption of the algebraic notation is if other mathematical formulae such as those for differentiation and Taylor's theorem hold. So mathematical generality is very important.

In contrast, De Morgan who wished to 'investigate the forms of thought involved in combination of relations'[42] restricted his use of relations to the traditional syllogistic

---

[42]See [**90**, p. 212].

modes. Peirce provides only two examples of syllogistic reasoning and these only at the very end of the paper. The sign of illation $\prec$ is introduced by Peirce to mean IS INCLUDED IN. The use of the inequality rather than equality is justified in a footnote: 'inclusion in is a wider concept than equality, and therefore logically a simpler one'. This is one of many developments away from Boolean algebra. For the operations of identity and inclusion the symbols = and < are used, with $\prec$ (called by Peirce the symbol of illation and meaning 'as small as' or 'is included in') being used for the copula 'is'. De Morgan's influence is felt here as Peirce cites 'Formal Logic' in connection with the 'less than' < notation in a footnote on page 367. Peirce's novel use of the illation sign lies in the combination of < with = simply to mean IS so that f $\prec$ m means EVERY FRENCHMAN IS A MAN without saying whether there are any other men or not, whereas f < m implies that THERE ARE MEN BESIDES FRENCHMEN. Such a notation could have been suggested by De Morgan's double spiculae '((' .[43] A footnote to the *American Journal of Mathematics*, 1881 version of *Linear Associative Algebra* by Peirce, adds to his father's definition of $B < A$ given as A IS A WHOLE OF WHICH B IS A PART, SO THAT ALL B IS A, the implication that SOME A IS NOT B, thus clearly distinguishing his own illation operation from that of inequality.

Peirce was aware that his own logic of relatives was not without complexity. On page 368 of *Logic of Relatives* he stated:

> We labor under the disadvantages that the multiplication is not generally commutative, that the inverse operations are usually indeterminative, and ... equations ... where the exponents are three or four deep, are exceedingly common. It is obvious, therefore, that this algebra is much less manageable than ordinary arithmetical algebra.

This did not unduly worry Peirce, who like his father was not primarily concerned with the utility of his calculus. As he wrote on page 359:

> I think there can be no doubt that a calculus, or art of drawing inference, based upon the notation I am to describe, would be perfectly possible and even practically useful in some difficult cases, and particularly in the investigation of logic. I regret that I am not in a situation to be able to perform this labor, but the account here given of

---

[43]We have already seen this notation in "On the Syllogism: II" [**36**]. Also see page 65 above.

the notation itself will afford the ground of a judgement concerning its probable utility.

Logic of Relatives then satisfies all of Hamilton's criteria for an algebraic theory on a practical, philological and theoretical level. The algebraic logic developed in *Logic of Relatives* is practical in that a list of definitions and rules of the basic operations as well as 173 theorems are set up, philological in that a notation and interpretation is given for a number of logical terms and theoretical since algebra is carefully included with the logic.

To answer the question "what are the axiomatic principles of this branch of logic?," let us note that Peirce first seems to identify an axiom as that which is 'not deducible from others' and then states without proof, that these are the general equations given under the heading of "Application of the Algebraic Signs to Logic", (probably §1 - §20 as these formulae are not listed formally until the next section under the heading of "General Formulae"), together with those relating to backward involutions, (approximately 20 of these), formulae §95, §96 concerning individual terms, §122, §142, §156 (the 4×4 multiplication table), §25, §26 (the laws of contradiction and excluded middle), §14, §15, §169, and §170 If $x < y^z$ then $z(Ky)^x$. However it is not clear which formula §169 is intended here as Peirce in fact gives two! One as quoted above dealing with the converse conjugative term and §169 on page 421 dealing with a property of 'not' or ᒎ, (where this is the relative term OTHER THAN _____), namely §169 If $[x] > \mathbf{1}, ᒎ^{-1}x = 1$.

So although Peirce numbers 172 formulae, he presents 173 with §169 being used twice for two different formulae. It is probable that the intended §169 is the formulae dealing with converses since §170 is also cited. Having developed 173 logical formulae and then selected from these approximately fifty 'independent' general formulae which he calls 'axioms', Peirce now writes 'But these axioms are mere substitutes for definitions of the universal logical relations, and so far as these can be defined, all axioms may be dispensed with.' In his conclusion to *Logic of Relatives*, Peirce takes the view that the fundamental principles of formal logic are not axioms but definitions and classifications, with their validity justified by analogy with familiar processes, i.e. mathematical processes: 'the only <u>facts</u> which it contains relate to the identity of the conceptions resulting from those processes with certain familiar ones'.

4.5.1. Early Writings in the Logic of Relations. We shall now consider Peirce's development of his logic of relations, which developed in parallel with his work on Boolean algebra. The early writings of Peirce on relations started from a consideration of logical categories greatly influenced by Kant and continued when he investigated further traditional syllogistic logic from ancient logicians up to De Morgan. Starting with a method of multiple subsumption (multiple occurrences of simultaneous substitution) Peirce used 'Rule, Case, Result' for dealing with syllogisms, e.g. in

> Whatever number results from the multiplication of one by another results also from the multiplication by that one of the other 12 results from the multiplication of 4 by 3 Therefore, 12 results from the multiplication by 3 of 4,

found in "Lowell Lecture II" (1866)[44], which, as Merrill points out, is similar to De Morgan's dictum de majore et minore[45]. Briefly this principle stated in *Formal Logic* [**33**] substitutes a lesser term for a universal term and a greater one for a particular term, as in the following example:

> Every head of a man is the head of a man.
> Man is an animal.
> Therefore, every head of a man is the head of an animal.

However, Peirce avoided the need for multiple subsumption by using a relational approach to the syllogism. Consider the following mathematical syllogism:

> Every part is less than that of which it is a part,
> Boston is a part of the Universe;
> Therefore Boston is less than the Universe.

This is then reduced to the syllogistic form:

> Any relation of part to whole is a relation of less to greater,
> The relation of Boston to the Universe is a relation of part to whole;

---

[44]See [**82**, pp. 376–392].
[45]See [**122**, p. 260].

Therefore the relation of Boston to the Universe is a relation of less to greater.

The whole area of quantification in logic was a very important one, and Peirce was to write in 1893 that the further study of this subject gave him the whole theory of the logic of relatives (MS 811). Although as we have seen, consideration of the importance of relations occurred at an early date, i.e. 1867, we will now consider several links between Peirce's notation for relatives and De Morgan's "On the Syllogism: IV." The first experiments with a notation began in November 1868. Merrill has stated that in Peirce's 'Logic Notebook', during the period November 3-15, 1868, Peirce used a subscript notation for quantifying relations, using l_W to represent both LOVER OF SOME WOMAN and LOVER OF EVERY WOMAN. De Morgan was also to consider this distinction in his contrary relations letting $X.LMY$ stand for X IS NOT ANY L OF ANY M OF Y rather than X IS NOT ANY L OF SOME OF THE Ms OF Y in his "On the Syllogism: IV." Merrill shows that in one additional (unpublished) page of the Notebook of uncertain date Peirce compares his notation with that of De Morgan. Peirce's notation used here was not that of *Logic of Relatives*, e.g., he represented De Morgan's LM' as $v(1-(1-L)M)$ rather than l^m. The use of v the indeterminate class of Boole also shows this was an early work as Peirce deliberately avoided this symbol in *Logic of Relatives* referring to it as a Boolean symbol, using the alternative form 0/0 for the one instance where it occurs in an example of his. This shows that De Morgan's superscript notation of LM' was in his mind rather than $1 - (1 - L)M$ the Boolean form when formulating l^m.

This note also gives details of De Morgan's subscript notation L, M denoting AN L OF NONE BUT Ms which Peirce is later to describe as 'backwards involution' in *Logic of Relatives*. However in this note the *Logic of Relatives* notation of $^l m$ is not used by Peirce, but an earlier Boolean influenced notation of $v(1 - L(1 - M))$. If this note was written before *Logic of Relatives*, it shows that Peirce was aware of De Morgan's backwards involution at this early stage. Peirce, however, later claimed that the inclusion of backwards involution was an afterthought prompted by reading De Morgan's "On the Syllogism: IV" [**40**] and that *Logic of Relatives* was almost complete before De Morgan's paper was read. Merrill has pointed out that Peirce's resistance to backwards involution could have arisen because it violates one of the defining conditions of algebraic exponentiation, i.e. $^z(^y x) = {}^{zy}x$ rather than $^z(x^y) = {}^{zy}x$, thus weakening any algebraic analogy with his system.[46]

---

[46]See [**122**, p. 279].

The question remains whether Peirce discovered the logic of relations independently of De Morgan. Emily Michael in her article "Peirce's Early Study of the Logic of Relations, 1865 - 1867"[47] has shown that Peirce's early work contained in other articles of this period led him to see the incompleteness of traditional syllogistic and so take relations into account. In these papers Peirce considered an extension of syllogistic logic to a consideration of dyadic relations with different types of relative terms and how to convert syllogistic forms to relational propositions. This involved a study of certain convertible (our modern-day term 'symmetrical') relations called equiparent (relations of agreement), and disquiparent relations (relations of opposition). Peirce seems to have come by these relations via Ockham rather than De Morgan.[48]

4.5.2. Comparison with Boolean Algebra and De Morgan's Logic of Relations.

It is probable that although Peirce's earliest work on the logic of relations was independent of "On the Syllogism: IV", De Morgan's influence inspired the initiation of *Logic of Relatives*. One example is in the use of Peirce's sign of illation $\prec$ as the fundamental logical relation rather than the Boolean =. Peirce states in a footnote on page 360 that $\prec$ is used in place of < because ≤ cannot be written rapidly enough. This is surprising in view of the fact that Peirce could have used De Morgan's spiculae '((' which would be even quicker to write. It seems that Peirce is deliberately avoiding the spiculae notation. A more important reason given was that Peirce considered the operations of equality and 'less than' as special cases of inclusion, which is therefore the broader and simpler logical concept. One of the main modifications of the Boolean calculus of classes was "inclusive" addition in which elements were not counted twice and so the equation $x + x = x$ holds (as proposed by Jevons). Peirce also introduced relative sums for the first time, which De Morgan did not consider, but this definition seems straightforwardly analogous to addition between classes (or absolute terms). A further development away from Boolean algebra was the use of the inclusion sign as the sign for the fundamental logical relation rather than Boole's equality sign =. It should be noted however that of the axioms presented, only 27 of them use $\prec$ rather than =. Taken together with De Morgan's spiculae, this move away from equations seems to prepare the way for a logic where implications rather than equations are used.

Another logical influence was Sir William Hamilton the foremost logician in Britain at the time. We have already seen in "Harvard Lecture VI" that Peirce was

---

[47]See [**126**, pp. 63–67].
[48]See [**55**, p. 334].

well aware of Hamilton. He described in this paper Hamilton's syllogistic notation and in "Harvard Lecture XI" he discussed ways of amending Sir William Hamilton's eight postulates. Michael has analyzed extensively Peirce's treatment of conditional arguments and also those involving relative terms in "Lowell Lecture II"[49]. However this paper seems to be influenced with the philosophy of De Morgan's *Formal Logic*, i.e. that mathematical reasoning can be brought under the traditional syllogistic system and in particular Peirce introduces here the concept of De Morgan's numerically definite syllogisms and provides a defence of it.

Michael and Merrill have shown that Peirce's own later recollections on the subject are inconsistent. Merrill states: "Sometimes he claims that his work on relations was essentially independent of De Morgan's, while at other times he says just the opposite."[50]. It seems clear that Peirce had read De Morgan's "On the Syllogism: IV" which introduces the logic of relations before he came to write *Logic of Relatives*. De Morgan wrote to Peirce in April 1868 promising to send "On the Syllogism: IV" and "On the Syllogism: V" and Peirce refers to it in a paper published in 1869 "Grounds of Validity of the Laws of Logic."[51] He also delivered a lecture on De Morgan's "On the Syllogism: IV", in the month before *Logic of Relatives* was communicated to the American Academy of Arts and Sciences on January 26,1870. Although Michael has shown that Peirce may well have devised a logic of dyadic relations in 1866 prior to *Logic of Relatives* there is no doubt that "On the Syllogism: IV" was an inspiration for *Logic of Relatives*. Even "Note 4" [**147**] composed in Nov-Dec 1868, where the rudiments of *Logic of Relatives* are found uses a superscript notation for both application of relations and ordinary involution. Moreover the main example given is the De Morgan challenge EVERY MAN IS AN ANIMAL. Therefore, ANY HEAD OF A MAN IS A HEAD OF AN ANIMAL. The same example is also referred to in "Grounds of Validity of the Laws of Logic," probably in late December 1868, in a footnote:

> If any one will by ordinary syllogism prove that because every man is an animal, therefore every head of a man is a head of an animal, I shall be ready to set him another question.

Merrill writes in his introduction to [**83**]:

---

[49]See [**55**, pp. 376–392].
[50]See [**122**, p. 247].
[51]See [**56**, p. 245].

It is thus very likely that Peirce had read De Morgan's paper before he wrote the entries in LN dated November 1868, even though those entries carry no clear references to De Morgan and use quite different examples.

However we have seen that even in the *Logic Notebook* of 1868, "Note 4" carries a distinct reference through use of the De Morgan challenge, to De Morgan and even perhaps through the reference in "Grounds of Validity of the Laws of Logic" to "On the Syllogism: IV" itself. So that at the time when Peirce was sketching out the rudimentary forms to be developed in *Logic of Relatives* in "Note 4" of Dec 1868, it is clear that he used the head-of-a-man example provided by De Morgan, and it seems fairly likely, as hinted through the footnote to the "Grounds of Validity" paper sent to the printers in late December 1868, that Peirce was using this example from "On the Syllogism: IV" rather than from *Formal Logic*. Also note that this challenge refers to the application of a relative to an individual or class term. Furthermore, it is not true that all references by Peirce at the time that he started to study the logic of relations, refer to De Morgan's *Formal Logic* [**33**] as in his paper "Upon Logical Comprehension and Extension" of 1867[52] De Morgan's "Syllabus of a Proposed System of Logic" [**38**] is cited. Peirce himself states that he discovered the logic of relations independently:

> I was not the first discoverer, but I thought I was, and had complemented Boole's algebra so far as to render it adequate to all reasoning about dyadic relations, before Professor De Morgan sent me his epoch-making memoir in which he attacked the logic of relatives by another method in harmony with his own logical system.

The use of illation to represent the fundamental logical relation follows De Morgan's lead since De Morgan used, as Peirce puts it in his introduction to *Logic of Relatives*, his 'well-known spiculae' rather than the equality = of Boole. However as noted previously, only 27 of the 173 formulae use illation rather than equality. De Morgan's influence was more apparent in the section on 'backwards involution' since Peirce extensively revised *Logic of Relatives* to incorporate De Morgan's 'backward involution' given in "On the Syllogism: IV." Peirce also added the work on converse relatives at this time. Another break with De Morgan was the introduction of Peirce's conjugative terms which considered for the first time n-adic relations rather than the dyadic relations used by De Morgan. Merrill concludes: 'while Peirce probably knew of De

---

[52]See [**83**, p. 71].

Morgan's memoir on relations when he was working out the full notation of *Logic of Relatives*, his own Boolean orientation meant that he was working on these topics in his own way."[53]

It can be seen that few proofs of Peirce's formulae are given. Like De Morgan, Peirce was content with stating formulae and providing examples in the form of English sentences. He does however introduce each set of formulae with a discussion and sometimes sketches of proofs as we have seen above and also once provides a sketch of proofs of formulae §30–§33, Boole's development theorems, in a footnote. In general however, Peirce only justifies his formulae by reference to previous formulae leaving the reader to verify the proofs. Like De Morgan, he investigated contraries, converses, relative products and involution. However he also extends the logic of relations to include the universal and null relations and the identity relation as well as considering the logical sum of relations, triadic relations and relations of higher degree. Martin writes:[54]

> Peirce occasionally observes that one principle is a special case of another, or that "it is easy to show" such and such. But no acceptable proof in the modern sense is ever given in this paper. But note that all but a few of his formulae, and more general forms of them, are readily provable in the modern theory of (virtual) classes and relations as based on quantification theory with identity and abstraction. That Peirce was able to put forward acceptable and important principles independent of that theory, on the basis of his clumsy notation and inadequate deductive framework, is remarkable indeed and well attests to his extraordinary logical insight.

Throughout *Logic of Relatives* there is little mention of the syllogism. Whereas De Morgan and Boole took the Aristotelian syllogism as their foundation, Peirce has accepted De Morgan's position that it is inadequate and limiting. It is by incorporating the logic of relations through De Morgan's 'improvements' into the Boolean logic of classes, that Peirce can overcome the restrictions and inadequacies of syllogistic logic that he first noted in his "Harvard Lecture III" and "Harvard Lecture VI" in Boole's

---

[53]See [**123**, p. xlv].
[54]See [**121**, p. 44].

logic. This move from the revered syllogism quickly reached a pinnacle with Lewis commenting in 1918:[55]

> To regard the syllogism as indispensable, or as reasoning par excellence, is the apotheosis of stupidity.

However it was not always so. In early writings such as "Lowell Lecture II" influenced by the renewal of interest in logic in England culminating in De Morgan's series entitled "On the Syllogism," Peirce believed that traditional syllogistic logic could be expanded to encompass all mathematical inferences. He wrote "mathematical demonstrations can be reduced to syllogism" and gave this as a reason for attaching greater importance to logical studies.[56] However, having seen the limitations of the syllogism he was later to claim that logic is part of mathematics. Diepert has the following:[57]

> But Peirce meant by mathematics something more like the systematic and rigorous theory of diagrams and formal representations used in necessary reasoning. Mathematics would thus include not only formulas, diagrams, graphs and so on that mathematicians do employ, but also, for example, the grammar and transformation rules of natural languages.

He also differed from De Morgan, in that he considered the application of classes of relatives to absolute terms, while De Morgan studied mainly the application of relations to relations and to a lesser extent, relations applied to individuals or generic members. This emphasis on classes is natural given Peirce's avowed aim given in the title of *Logic of Relatives*, namely of amplifying the conceptions of Boole's calculus of logic. The titles of these two works also contrast accurately Peirce's development of relative terms as in "Description of a Notation for the Logic of Relatives," while De Morgan's full title of his 1860 paper is "On the Syllogism: IV" and "On the Logic of Relations." Peirce uses ordinary involution (exponentiation) with classes, as opposed to De Morgan who used involution only with relations. Throughout *Logic of Relatives*, he is seeking to extend both the Boolean algebra of classes and De Morgan's logic of relations to a wider mathematical context involving both a form of differentiation and

---

[55]See [**115**, p. 2].
[56]See [**82**, p. 386].
[57]See [**48**, p. 46].

Taylor's theorem as well as quaternions and a geometrical interpretation. While De Morgan provided the inspiration for *Logic of Relatives*, it was a Boolean approach in terms of the use of equations and classes rather than De Morgan's abstract copula of inclusion and his relations that influenced its development.

Apart from the logical influences of Boole and De Morgan, a purely mathematical influence (probably a mutual one, since Peirce later claimed in a letter to Frederick Adams Woods dated 11 September 1913 that *Linear Associative Algebra* was a research that his father would never have undertaken but for 'my constantly pestering him to do so') came from Benjamin Peirce's *Linear Associative Algebra* published in the same year as *Logic of Relatives*, i.e. 1870. In *Logic of Relatives*, through his elementary relatives[58] Peirce created a logical interpretation for the linear algebras produced in *Linear Associative Algebra*. Furthermore his elementary relatives provide a linear representation for such matrix algebras. In this he later discovered he had been anticipated by Arthur Cayley. J. J. Sylvester also used similar forms and when Peirce placed a sentence of his own while proof reading Sylvester's paper on nonions which read "These forms can be derived from an algebra given by Mr. Charles S. Peirce."; this led to an angry dispute between the two.[59]

Peirce tentatively suggests that all such linear algebras can be expressed in the form of his elementary relatives. In 1881, in the American Journal of Mathematics, vol. 4, pp. 221–229, Peirce published notes on his father's *Linear Associative Algebra* showing the relationship of these algebras to the logic of relatives. Entitled "On the Relative Forms of the Algebras" they formed the second part to the Addenda published in the *American Journal of Mathematics*, of which the first part "On the Uses and Transformations of Linear Algebra" by Benjamin Peirce was presented before the American Academy of Arts and Sciences.[60] In this latter paper, a number of applications and 'uses' of linear associative algebras are presented but the brevity of the arguments are disappointing in comparison with the research already carried out in calculating the multiplication tables for the algebras. This slim paper was to take the place of the two promised volumes that Benjamin Peirce mentioned in *Linear Associative Algebra*. However, as we have seen, such applications were not his primary concern. One of the most important uses cited was that all linear associative algebras could be expressed by Charles Peirce's logic of relatives.

---

[58]See Section 4.2.4 above.

[59]See [**17**, p. 140].

[60]See [**138**].

Peirce first showed this in his paper "On the Application of Logical Analysis to Multiple Algebra" [**150**], although it was stated as very likely true in *Logic of Relatives* (on the grounds of inductive evidence). The 1875 paper takes for each algebra an 'absolute algebra', i.e. a linear associative algebra, whose general expression is a linear combination of its units, $aI + bJ + cK + dL + \ldots$ This represents the product of a multiplication and cannot be a multiplier. Then a unit i of the corresponding relative algebra acting on this, also gives on multiplication, a linear combination of I, J, K. ... That relative can then be expanded into a linear combination of relatives of the form $aA : B$, a being a scalar, such that $(A : B)(B : C) = A : C$ and $(A : B)(C : D) = 0$ and the product of these with any unit of the absolute algebra is another letter of the algebra.

In the 1881 Addendum which is a restatement of his 1875 paper "On the Application of Logical Analysis to Multiple Algebra," Charles Peirce defined the relative form of a linear associative algebra in the following way:

> Given an associative algebra with units I, J, K and L then define new units A, I, J, K, L where I, J, K and L correspond to the units of the algebra. These units can be multiplied by numerical coefficients and added but they cannot be multiplied together, and so are called nonrelative units.

The following properties hold for operations defined to be of the form $(I : J)$

(1) $(I : J)(aI + bJ + cK) = bI$
(2) $(J : K)K = J, (K : L)L = K$
(3) $(I : J)(J : K) = (I : K)$
(4) $(I : J)(K : L) = 0$

In particular (1) is explained by Peirce:

> Any one of these operations performed upon a polynomial in non-relative units, of which one term is a numerical multiple of the letter following the colon, gives the same multiple of the letter preceding the colon.

(2), (3) and (4) follow from (1). Peirce reasoned that (3) follows: since $(J : K)K = J$ and $(I : J)J = J$ so $(I : J)(J : K)K = (I : J)J = I$ so that $(I : J)(J : K)$ must be $(I : K)$. In a similar proof for (4) Peirce wrote:

$(I : J)(K : L) = 0$; for $(K : L)L = K$ and $(I : J)K = (I : J)(0.J+K) = 0.J = 0$

However Peirce omits the crucial step:

Therefore $(I : J)(K : L)L = (I : J)K = 0$ so $(I : J)(K : L)$ must be 0.

Just after this statement, the distributive law is assumed so that

$\{(I : J) + (K : J) + (K : L)\}(aJ + bL) = aJ + (a + b)K$

Of course, Peirce intends $\{(I : J) + (K : J) + (K : L)\}(aJ + bL) = aI + (a + b)K$ here, since $\{(I : J) + (K : J) + (K : L)\}(aJ + bL) = aI + aK + bK$. This error is not recorded in the 1933 Hartshorne & Weiss edition.

Complex operations are then defined which consist of a linear combination of operations of the form $(I : J)$ but with the addition of one operation $(I : A)$ for i', $(J : A)$ for j', etc. These complex operations i', j' etc. which take the form of relations between A, I, J, K etc. are shown to be equivalent to the original units of the algebra i, j, k etc. in the sense that their multiplication tables are equal. The method is to take $i'j'A = k'l'A$ and show that $i'j'M = k'l'M$ for any of the original units, so that the multiplication tables for i', j', etc. will be the same as that for i, j, etc. It is interesting to note that Peirce never refers to the letters or units of the algebra as *vids* which his father proudly proclaimed was the name that his son had devised and which was used in his own section of the addenda in the same article. The article concludes with a statement of complex numbers and quaternions in relative form. Complex numbers being represented by $1 = (X : X) + (Y : Y)$, $J = (X : Y) - (Y : X)$ and quaternions by:

$1 = (W : W) + (X : X) + (Y : Y) + (Z : Z)$
$i = (X : W) - (W : X) + (Z : Y) - (Y : Z)$
$j = (Y : W) - (Z : X) - (W : Y) + (X : Z)$
$k = (Z : W) + (Y : X) - (X : Y) - (W : Z)$

Peirce stated that the proof given here is essentially the same as that given in the 1875 paper, but in the former paper he has clarified matters by introducing an extra unit A into the expansion of his relatives and then by showing that the multiplication of such relatives is equivalent to the multiplication of the original algebra.

CHAPTER 5

The Theory of Quantification as Introduced by C. S. Peirce in his Later Papers on the Algebra of Logic

5.1. *Logic of Relatives* Onwards. Innovations in Later Peirce Papers up to 1883

5.1.1. Developments in Peirce's Logic after 1870. Surprisingly, after *Logic of Relatives* very little is attempted by Peirce in developing his algebraic logic until the early 1880s, when he introduces his theory of quantification. How do we account for this gap? A clue appears in the introduction to [**84**] where Fisch writes:

> There was no more intensively scientific seven-year period of Peirce's life than that of the present volume. He had no academic employment and gave no lectures at Harvard or at the Lowell Institute or elsewhere. As an Assistant in the Coast Survey his duties had so far been astronomical, and his concurrent assistantship in the Harvard College Observatory (1869–72) had been arranged with a view to those duties. But from late in 1872 onward his duties became increasingly geodetic.

Parental pressure was also brought to bear when Benjamin Peirce, who supervised his work with the American Coast Survey, advised his son not to make a career from logic but to continue with science. This advice was given on the occasion of Benjamin Mills Peirce's death in 1870. Charles' younger brother, a mining engineer by profession but also a talented artist, had led a frenetic but undisciplined life and died young. Recognizing the same weaknesses in Charles, Benjamin wanted him to remain in a profession that gave him a secure income, i.e. to continue with the Coast Survey. Also during this time, Charles became increasingly involved with philosophy. Stimulated by the meetings of the Metaphysical Club, founded together with Cambridge friends such as Chauncey Wright and William James, he developed his ideas by presenting and discussing papers on philosophical issues. This led to the birth of the philosophy of pragmatism as developed by James and Peirce, who called his own version pragmaticism.

5.1.2. The Relation between Mathematics and Logic in later Logic work.
Some work on further applications of his algebraic logic began in 1873, but it was not until 1880 that Peirce published his results as an *American Journal of Mathematics* article "On the Algebra of Logic" [**153**]. His more ambitious dream of writing a book on his life's work on algebraic logic was never realized, although a rough draft for this exists. More a summary of his logical developments than a completely new notation and methodology as *Logic of Relatives* had been, the early drafts he wrote for the proposed book on logic in 1873 show a great contrast in approach from *Logic of Relatives*, where the Boolean philosophy of applying mathematical techniques to logic was predominant. In these early years, Peirce's philosophical position emphasized the importance of logic, claiming that algebra is part of logic. This can be seen when he wrote in MS 221, March 14,1873, a draft of Chapter 7 entitled "Of Logic as a Study of Signs":[1]

> The business of Algebra in its most general signification is to exhibit the manner of tracing the consequences of supposing that certain signs are subject to certain laws. And it is therefore to be regarded as part of Logic.

In the same work, he defined logic as the science of identity and mathematics as the science of equality. Furthermore, mathematics was for Peirce the logic of quantity, allocating mathematics firmly as part of logic.

However Peirce's views on the relationship between logic and mathematics proved to be constantly changing and six years later in 1879, he had taken up yet another position. He now stated that mathematics and logic are distinct subjects. He wrote in "On the Algebraic Principles of Formal Logic," a work which is a fragmentary sketch of a systematic treatment of algebraic logic, that:

---

[1]See [**84**, pp. 82–83].

[t]he effort to trace analogies between ordinary or other algebra and formal logic has been of the greatest service; but there has been on the part of Boole and also of myself a straining after analogies of this kind with a neglect of the differences between the two algebras, which must be corrected, not by denying any of the resemblances which have been found, but by recognizing relations of contrast between the two subjects.

Peirce frequently contrasted the mathematical and the logical interest in notations. He claimed that "the mathematician's aim is to facilitate calculation, inference, and demonstration; the logician's, to facilitate the analysis of reasoning into its minimal steps."

By 1885 in "On the Algebra of Logic, A Contribution to The Philosophy of Notation" [**158**], his position seemed almost completely opposite to that taken in 1873. He now denied the very algebraic notation that provided his initial inspiration and claimed that logic should be pre-eminent. He wrote:

Besides, the whole system of importing arithmetic into the subject is artificial ... The algebra of logic should be self-developed, and arithmetic should spring out of logic instead of reverting to it.

In fact he claimed that it was to be through logic that new methods of discovery in mathematics would be found.

5.2. Major Innovations After *Logic of Relatives*

There are three main innovations arising out of Peirce's further work on logic in the decade after *Logic of Relatives* was published. These are duality, a modal logic system, and a fourth logical operation called 'transaddition'.

5.2.1. Duality. The emphasis on duality was largely absent from *Logic of Relatives*. One example from his 1879 paper, "On the Algebraic Principles of Formal Logic" mentioned above, is the following duality theorem:

THEOREM 1. *Corresponding to every general proposition of logic deducible from*

$$If\ x \prec y\ and\ y \prec z\ then\ x \prec z$$

without taking into account any other character of the copula, there is a proposition obtainable from the first by everywhere interchanging $\prec$ with $\succ$.

This theorem may have been inspired by Schröder, since the earlier part of (Peirce 1879) contains many examples of Schröder's formulae. Duality is again evident in the 1880 paper, "On the Algebra of Logic." This logic paper which was probably prepared for Peirce's lectures at the Johns Hopkins University where he was a part-time lecturer, was published in the *American Journal of Mathematics*, and formed an extended version of "On the Algebraic Principles of Formal Logic." In this paper Peirce conceived of a term as either an infinite logical sum of individuals or alternatively a negative term, which Peirce called a *simple*, as an infinite logical product of negatives. It can be seen that a negative term is what we would now use as a complement in set theory. In particular the definitions of addition and multiplication of Boolean terms, which Peirce calls *non-relative* are given in terms of dual formulæ:

$$a \prec x\ \&\ b \prec x \Leftrightarrow a + b \prec x$$

and

$$x \prec a\ \& x\ \prec b \Leftrightarrow x \prec a \times b$$

This is the first such definition of the operations which previously had been taken as aggregates of individuals by Boole (counting common terms twice) or by Jevons and Peirce (counting common terms once). However addition of relative terms is not explicitly defined, and seems to be a straightforward aggregation of relative terms (common terms counted once). As he stated in "On the Algebra of Logic":[2]

> The negative formulae are derived from the affirmative by simply drawing or erasing lines over the whole of each member of every equation.

_____

[2]See [**153**, p. 208].

He further elaborated this in "On the Algebra of Logic, A Contribution to the Philosophy of Notation" [**158**] in the following way, showing clearly his debt to Schröder:

$$\overline{x + y} = \overline{x}\,\overline{y}$$

$$\overline{x} + \overline{y} = \overline{xy}$$

"The apparent balance between the two sets of theorems exhibited so strikingly by Schröder, arises entirely from this double way of writing everything."[3]

5.2.2. Modal Logic. Another of the innovations introduced in [**153**] is the introduction of a new notation for a bi-valued logic, i.e. one with variables **v** and **f** for true and false respectively. The first sections of this work treat the syllogism by considering a valued logic where P is the class of all premises and C is the class of all conclusions, so that Pi ≺ Ci means that every state of things in which a proposition of one of the classes of premises is true is a state of things in which the corresponding propositions of the class Ci are true. A quote from "On the Algebra of Logic":

> Logic supposes inferences not only to be drawn, but also to be subjected to criticism; and therefore we not only require the form P therefore C to express an argument, but also a form, Pi ≺ Ci, to express the truth of its leading principle. Here Pi denotes any one of the class of premisses, and Ci the corresponding conclusion. The symbol ≺ is the copula, and signifies primarily that every state of things in which a proposition of the class Pi is true is a state of things in which the corresponding propositions of the class Ci are true. However I should add that Pi ≺ Ci also implies either 1, that it is impossible that a premise of the class Pi should be true or 2, that every state of things in which Pi is true is a state of things in which the corresponding Ci is true.

In comparison with *Logic of Relatives*, other new features include the overstrike bar to indicate the class complement and the symbol ∞ used instead of 1 for the universe. Peirce also calls ∞ and 0 the terms for the possible and the impossible. The logical terms called absolute, relative and conjugative are redefined. A relative is now a term which describes the class of *relates* of the relation so that as in *Logic of Relatives*, A : B defines the relation which has domain A and range B. However it must be

---

[3]See CP 3.386 in [**158**].

noted that $A : B$ is not the relation but rather is the class A which has B as the correlate. The elementary relative of *Logic of Relatives* in which individuals are related is now called a *dual relative*. Absolute terms are now called terms of singular reference. By adding an indefinite term to the system, these terms of singular reference may be written

(1) $$A = A : A + A : B + A : C + \dots$$

and

(2) $$B = B : A + B : B + B : C + \dots$$

We also have

(3) $\qquad A : B = A : (B : A) + A : (B : B) + A : (B : C) + A : (B : D) + \dots$

Peirce then states that where the relation is *coexist* we have,

(4) $\qquad A : B = (A : A) : B + (A : B) : B + (A : C) : B + (A : D)B + \dots$

Comparing this with (3), apparently the associative law is contradicted. But it is to be noted that (1) is the expression of a term of single reference as an infinite dual relative by means of the relation *coexisting with*. It is because this relation is commutative that there is in fact no contradiction. Since the relation is associative, writing (4) as

(5) $\qquad A : B = A : (A : B) + A : (B : B) + A : (C : B) + A : (D : B) + \dots$

we can see, since the relation is commutative that (4) is

(5) $\qquad A : B = A : (B : A) + A : (B : B) + A : (B : C) + A : (B : D) + \dots$

which is (3).

In addition to the three operations of relative multiplication, forwards involution and backwards involution, all inspired by De Morgan, Peirce now added a fourth operation which he named *transaddition*. If ls denotes WHATEVER IS A LOVER OF A SERVANT then $l\dagger s$ denotes $\overline{\bar{l}\bar{s}}$ or 'whatever is not a lover of everything but servants'. In other words the negative of relative multiplication.

This remarkable operation is the first instance of a relative sum, however it must be noted that this operation is not to be confused with that of class union or disjunction. Peirce had originally defined relative sum as

$$a \circ e = \bar{\bar{a}} \ \bar{\bar{e}}$$

in "On the Algebra of Logic.". It is in "Note B" that he changes it to the definition given above.

5.2.3. Peirce's 1880 Paper "A Boolian Algebra with One Constant".

Written in the winter of 1880, this paper shows the remarkable innovative power of his thought. Peirce's main interest in this work was in reducing the, number of logical operations to one, not counting colons, semicolons, periods and parentheses used as a means of separation. Here A means that the proposition A is true, and AA means that A is false and AB that both A and B are false. The proposition "If S (is true), then P (is true)" is expressed as $SS, P; SS, P$ which means that SS, P is false, but SS, P means "If S is false then P is true" or "If S is true then P is false" and the negative of this is then 'If S, then P'.

It should be noted that Peirce's method of repeating the logical variable as a sign of negation (and therefore complementation) implies that the only logical operation needed is that of taking the complement. Peirce seems to have anticipated the later development of the Sheffer stroke in propositional calculus. Peirce wrote on page 221:

> Of course, it is not maintained that this notation is convenient; but only that it shows for the first time the possibility of writing both universal and particular propositions with but one copula which serves at the same time as the only sign for compounding terms ...

However this was qualified by Peirce's footnote to his student Christine Ladd's paper also entitled "On the Algebra of Logic"[4], where he stated:

> Every algebra of logic requires two copulas, one to express propositions of non-existence, the other to express propositions of existence.

---

[4]See [**110**, p. 23].

This corresponds more closely to his later position as shown in (Peirce 1896), where he held that the primary and fundamental logical relation was that of illation, expressed by *ergo*. It seems that because of this position, Peirce did not further advance his development of the *nand* operation. Zeman notes:[5]

> Peirce shows that "neither-nor" is a sufficient sole connective for the classical propositional logic; this is thirty-three years before Sheffer's showing and being acclaimed for showing that one such connective can suffice.

By 1880, in "On the Algebra of Logic," Peirce had defined the table of sixteen forms of the logical binary connectives for the first time in a matrix formation after the style of his father Benjamin Peirce's linear algebras. When considering the logic of two propositions X and Y, there are sixteen possible relations between these propositions such as X AND Y, X OR Y, etc. These relations are often called the sixteen binary connectives. First considered by De Morgan, it can be seen that using three individuals and two relations, sixteen possible propositions can be formed, namely:

$$(A{:}B)(B{:}C) \quad (\overline{A{:}B})(B{:}C) \quad (A{:}B)(\overline{B{:}C}) \quad (\overline{A{:}B})(\overline{B{:}C})$$
$$(A{:}B)(C{:}B) \quad (\overline{A{:}B})(C{:}B) \quad (A{:}B)(\overline{C{:}B}) \quad (\overline{A{:}B})(\overline{C{:}B})$$
$$(B{:}A)(B{:}C) \quad (\overline{B{:}A})(B{:}C) \quad (B{:}A)(\overline{B{:}C}) \quad (\overline{B{:}A})(\overline{B{:}C})$$
$$(B{:}A)(C{:}B) \quad (\overline{B{:}A})(C{:}B) \quad (B{:}A)(\overline{C{:}B}) \quad (\overline{B{:}A})(\overline{C{:}B})$$

It seems that here as so often elsewhere Peirce was a remarkable innovator. Shea Zellweger writes:[6]

> The logic of propositions is a fundamental part of symbolic logic. If one gives central emphasis to the role of symmetry, when great care is put on shape designing what it takes to construct a special set of sixteen iconic signs, then it is possible to bring to the logic of propositions an approach that not only simplifies and consolidates.

---

[5]See [**206**, p. 8].
[6]See [**202**, p. 76].

This approach, with its emphasis on symmetry, also receives major assistance from the algebra of abstract groups ... It has practical implications for digital design, mirror logic, and optical computers.

New notations for many of the operations and their converses were introduced. For example Peirce defined a:b as the operation of putting A in place of B in the triple relative b, and defined the following operations of transposition on page 198. This is interesting because Peirce makes a minor error here.

$$
\begin{array}{rcl}
I & = & a{:}b + b{:}a + c{:}c \\
J & = & a{:}a + b{:}c + c{:}b \\
K & = & a{:}c + b{:}b + c{:}a \\
L & = & a{:}b + b{:}c + c{:}a \\
M & = & a{:}c + b{:}a + c{:}b \\
1 & = & a{:}a + b{:}b + c{:}c
\end{array}
$$

He stated I+J+K=1+L+M, however this is not valid unless the operations of L and M above are interchanged, so that L should be defined as a:c + b:a +c:b and M as a:b + b:c + c:a.

5.3. Problem Solving and Applications by Peirce and his Pupils

In 1883, Charles Peirce published a volume entitled *Studies in Logic by Members of the Johns Hopkins University* [120]. This contained work by himself and his pupils Oscar Howard Mitchell, Christine Ladd-Franklin, Allan Marquand, and B. I. Gilman (who contributed a paper extending the logic of relatives to number and applying it to the theory of probabilities). I will now analyze the concepts and methods used by Ladd-Franklin, Mitchell, and in Peirce's own "Note B" from this volume, in particular concentrating on how they attempted to use their different versions of algebraic logic for problem solving.

5.3.1. Biographical Details of Christine Ladd-Franklin. Christine Ladd-Franklin was born on December 1,1847, in Windsor, Connecticut. Her ancestors were

prominent in Connecticut and New Hampshire.[7] Christine's father was a merchant; her mother died when she was thirteen years old. In her childhood, Christine dreamed of an academic education - something that was not readily available to the women of that time. However such was her determination and intelligence that she proved to be remarkably successful in achieving her goal. From the ages of twelve to sixteen, she attended school in Portsmouth. Then she was a student at Wesleyan Academy in Massachusetts for two years. Her studies included two years of Greek, a subject in which she was the only female student (Green 1987,122). She studied at Vassar College in 1866-1867. The lack of funds, however, prevented her return to Vassar. Instead, she taught one semester in Utica, New York, while studying trigonometry as well as the piano, biology, and several foreign languages. She also published an English translation of Schiller's "Des Mädchens Klage."[8]. Her fluency in German was to be important in her later correspondence and understanding of Schröder. In 1868 she returned to Vassar to continue her studies in languages, physics, and astronomy, but relatively little mathematics. However, by the time she graduated and returned to teaching, she was determined to learn more mathematics. The study of physics strongly aroused her intellectual enthusiasm, but Christine turned to mathematics as an area in which she could both pursue independent study and also develop her scientific creativity. While teaching in Washington, Pennsylvania, in 1871, Ladd-Franklin began contributing to the 'Mathematical Questions' section of the *Educational Times.*[9] She continued her study of mathematics at Harvard during the following year, under W. E. Byerly and James Mills Peirce (Charles Peirce's brother). By 1878 she had published several articles in the new American journal, *The Analyst,* as well as at least twenty mathematical questions or solutions to questions in the *Educational Times.* In that year she applied for admission to the graduate programme at the Johns Hopkins University even though the university was not open to women. J. J. Sylvester, who knew of her contributions to the *Educational Times,* urged that she should be admitted on a special status and granted a fellowship.

While at Johns Hopkins, Ladd came under the influence of her great mentor Charles Peirce. During this time she published three papers in the *American Journal of Mathematics* and wrote a dissertation in the area of symbolic logic. However, as Johns Hopkins would not award degrees to women, she left in 1882 without the degree of PhD. On August 24,1882, she married Fabian Franklin, a fellow student on

---

[7]See [**76**, p. 121].
[8]See [**101**, p. 354].
[9]See [**76**, p. 122].

the graduate programme and later member of the Johns Hopkins Mathematics Faculty. Even though she did not receive a degree, her dissertation was published in *Studies in Logic by Members of the Johns Hopkins University* [**110**].

Although continuing to work in symbolic logic, she also began investigations in the field of physiological optics. She published many papers in the field of color vision which was another of Charles Peirce's many interests. In 1892 she discussed her theory of color vision at the International Congress of Psychology in London[10]. She continued publishing on that subject during the next thirty-seven years, keeping a frequent correspondence with Peirce. Her collected works on color vision, *Colour and Colour Theories*, was published when she was eighty-one years old.

Christine was also an associate editor for logic and psychology for the 1902 *Dictionary of Philosophy and Psychology*, and she contributed many articles including two on logic co-authored with Peirce and letters to various newspapers and magazines. It was during this period that she was able to provide some assistance to her former supervisor Charles Peirce, who was at this time living in great penury, by soliciting many reviews and articles from him.

Christine Ladd-Franklin worked hard during her life in mathematics and science, but she was also remarkable in other ways. She spent much time, and some of her own money, helping women obtain a graduate education. She was awarded an LL. D. in 1887 by Vassar College, the only honorary degree that college has ever bestowed, and was finally granted a doctorate from Johns Hopkins University at the age of seventy-eight, forty-four years after the completion of her dissertation. On March 5,1930, she died of pneumonia at the age of eighty-two.

5.3.2. Ladd-Frankline's "On the Algebra of Logic" of 1883.

Ladd-Franklin's algebra in "On the Algebra of Logic" is primarily concerned with the syllogistic reasoning of Boole and is closer to Jevons's algebraic logic. The algebra uses the identity copula 'is' rather than with the relational logic of Peirce as described in *Logic of Relatives*. The paper concentrates on method and problem solving within the syllogistic framework, rather than notation and ignores relational logic apart from the traditional identity copula. Its main operation $\overline{\vee}$ as in A $\overline{\vee}$ B is equivalent to the class statement A ∩ B=0, otherwise the addition operation of Jevons and Peirce is used.

---

[10]See [**164**].

Logical multiplication and addition are defined clearly: a × b as the class of what is common to the classes a and b. When relative terms are excluded this may be written as ab, reserving the symbol (×) for arithmetical multiplication where necessary. Logical addition is defined as a+b the class of what is either a or b, where it takes in the whole of a together with the whole of b, what is common to both being counted once only.

What are Ladd-Franklin's classes? They are classes of individuals but more usually they are classes of logical propositions or predicates. However she also allowed the option that a ≺ b which means a is contained in b, where a AND b may be either terms or propositions. In general, upper case letters are used to represent predicates rather than propositions. She also provided a very clear definition of ∞, in contrast to Peirce, as:

> ... the universe of discourse, the universe of conceivable things or of actual things, or any limited portion of either ... In any proposition of formal logic, ∞ represents what is logically possible.

Again in contrast to Peirce, 0 is defined as the negative of ∞. She introduced two new operations as well as the standard addition and multiplication. She used AVB to mean A is partly in B and A $\overline{V}$ B to mean A is excluded from B or A is NOT B, so that x$\overline{V}$∞ means that x does not, under any circumstances exist, and xV∞ means that x is at least sometimes existent.

Ladd-Franklin then dispensed with the symbol ∞ so that x$\overline{V}$ means THERE IS NO x. In this way non-existence is clearly defined and Ladd-Franklin is then free to use propositions with her copula in the following way:[11]

> If a is a proposition, a$\overline{V}$ states that the proposition is not true in the universe of discourse. For several propositions, abc$\overline{V}$ means that they are not all at the same time true, so that a$\overline{V}$b means that propositions a and b are not both true at the same time.

Existence is also clarified: a$\overline{V}$b denotes either that the two propositions are logically consistent, or that they are possibly co-existent, or that they have actually been at some moment of time both true. Here truth and existence seem to be equivalent; truth being

---

[11]See [**110**, p. 30].

used for when the terms are propositions and existence for when they are used for predicates. This use of time reminds us of the idea of the time for which a proposition is true which comes from Boole.[12]

The key to problem solving in Ladd-Franklin's algebraic logic comes from the rule that the factors of a combination may be written in any order and the copula may be inserted at any position or it may be written at either end. This follows from

(17') $a\overline{\overline{V}}b = ab\overline{\overline{V}}$ and $abc\overline{\overline{V}} = a\overline{\overline{V}}bc = ca\overline{\overline{V}}b = \ldots$ and the proposition $abc\overline{V}de$ may be read "abc IS-NOT de", "cd IS-NOT abe", dc," and "abe IS-NOT,–that is IS EITHER NOT d OR NOT c," etc.

I shall now look at some specific examples of problem solving with Ladd-Franklin's notation and method, beginning with syllogistic expressions and finishing with an *Educational Times* problem of 1881. These examples are analyzed in some detail and I also show that in the final example, Ladd-Franklin has made a slight error.

Let us start with an amusing example of syllogistic propositions given by Ladd-Franklin on page 34 and 35:

$$(a\,\overline{V}\,b)\,(c\,\overline{V}\,d)\,\overline{V}\,(ac\,\overline{V}\,b{+}d).$$

This is interpreted as

IF NO BANKERS ARE POOR AND NO LAWYERS ARE HONEST, IT IS IMPOSSIBLE THAT LAWYERS WHO ARE BANKERS SHOULD BE EITHER POOR OR HONEST.

An alternative interpretation could be

IF CULTURE IS NEVER FOUND IN BUSINESS MEN NOR RESPECTABILITY AMONG ARTISTS, THEN IT IS IMPOSSIBLE THAT CULTURED RESPECTABILITY SHOULD BE FOUND AMONG EITHER BUSINESS MEN OR ARTISTS.

---

[12]See p. 78 above.

We shall now look more closely at Ladd-Franklin's method which is to obtain valid conclusions by taking a product or sum of any or all of the terms on the left-hand side of the copula and a product or sum of any or all of the terms on the right-hand side of the copula. For universal propositions the method is to obtain as a conclusion, a universal proposition of the product of the coefficients of $\bar{\bar{x}}$ and x, i.e., all those terms NOT x and x, and the sum of those terms in the propositions not including x.

This process can be seen by considering the following example given by Venn in the newly published quarterly journal *Mind*:[13]

> The members of a board were all of them either bond-holders or share-holders, but no member was bond-holder and shareholder at once; and the bond-holders, as it happened, were all on the board. What is the relation between bond-holders and share-holders?

Ladd-Franklin's solution is the following: Put a=member of board, b=bond-holder, and c=share-holder. The premises are

$$a \,\bar{V}\, bc + \bar{b}\,\bar{c},$$

and

$$b \,\bar{V}\, a$$

Taking the product of the coefficient of a by that of $\bar{\bar{a}}$, we have

$$b(bc + \bar{\bar{b}}\,\bar{\bar{c}}) \,\bar{V}.$$

Since bb = b and $b\bar{\bar{b}}$=0 we have the conclusion $bc\bar{V}$ or NO BOND-HOLDERS ARE SHARE-HOLDERS. The advantage of using the exclusion copula $\bar{V}$ rather than equality =, is that the copula $\bar{V}$ may be inserted at any position to obtain a valid conclusion.

---

[13]See [**200**, p. 487].

Let us finally examine in some detail the algebra of Ladd-Franklin as employed in the following example taken from the *Educational Times*, 1 February, 1881, by W. B. Grove, BA:

> The members of a scientific society are divided into three sections, which are denoted by a, b, c. Every member must join one, at least, of these sections, subject to the following conditions: (1) Any one who is a member of a but not of b, of b but not of c, or of c but not of a, may deliver a lecture to the members if he has paid his subscription, but otherwise not; (2) one who is a member of a but not of c, of c but not of a, or of b but not of a, may exhibit an experiment to the members if he has paid his subscription, but otherwise not; but (3) every member must either deliver a lecture or perform an experiment annually before the other members. Find the least addition to these rules which will compel every member to pay his subscription or forfeit his membership, and explain the result.

Ladd-Franklin began by outlining the premises: Put x = MUST DELIVER A LECTURE, y = MUST PERFORM AN EXPERIMENT, and z= HAS PAID THE SUBSCRIPTION. Then

$$(a) \qquad \bar{a}\,\bar{b}\,\bar{c}\,\overline{\overline{V}}$$
$$(1) \qquad a\,\bar{b} + b\bar{c} + c\bar{a}\,\overline{\overline{V}}\, xz$$
$$(2) \qquad a\bar{c} + c\bar{a} + b\bar{a}\,\overline{\overline{V}}\, yz$$
$$(3) \qquad \bar{x}\,\bar{y}\,\overline{\overline{V}}$$

According to Ladd-Franklin's method since these are all universal propositions of the form ALL A IS B, they are expressed with the negative copula $\overline{\overline{V}}$ to indicate exclusion, rather than the positive copula $\overline{V}$ which means that A and B have members in common and which is reserved for particular propositions, e.g., SOME A IS B.

(a) is to be interpreted as THERE ARE NO NON-MEMBERS OF A AND B AND C or the intersection of a, b and c is non-trivial so that there is at least a member in one of a, b or c. (1), (2), and (3) correspond to the original premises. It is required to eliminate z or those members who do not pay their subscription. Here (1), (2) may be written as

$$(a\overline{b} + bc + c\overline{a}) \times \overline{\overline{V}} \ z$$
$$(a\overline{c} + c\overline{a} + b\overline{a}) \ y \ \overline{\overline{V}} \ z$$

To eliminate z - THOSE WHO DO NOT PAY THEIR SUBSCRIPTIONS: these defaulting members represented by z do not appear on the left-hand side of 1) or 2) and since premises a) and 3) are universal exclusions, z needs to be eliminated from 'all that part of the universe from which it has not already been excluded; namely from the negative of

$$(a \ \overline{b} + b \ \overline{c} + c \ \overline{a})x + (a\overline{c} + c \ \overline{a} + b\overline{a})y + abc + xy$$

Ladd-Franklin states concisely that the negative of this is

$$(\overline{a} \ \overline{b} \ \overline{c} + abc + \overline{x})(\overline{a} \ \overline{b} \ \overline{c} + ac + \overline{y})(a + b + c)(x + y).$$

Continuing the analysis more closely using formulae first given in [37],

(13') $\overline{ab} = a + b$

and

(13°) $\overline{\overline{a} \ \overline{b}} = a + b,$

the negative of

$$(a\overline{b} + b\overline{c} + c\overline{a})x$$

is

$$(\overline{a} + b)(\overline{b} + c)(\overline{c} + a) + x$$

Multiplying out the last two factors of the product we obtain:

$$(\overline{a} + b)(\overline{b} \ \overline{c} + a\overline{b} + ac) + \overline{x}.$$

Using bb = b and b$\overline{b}$=0, this becomes

$$\overline{\overline{a}\overline{b}\overline{c}} + abc + \overline{x}.$$

Similarly the negative of $(a\,\overline{c} + c\overline{a} + b\,\overline{a})y$ is

$$\overline{a}\ \overline{b}\ \overline{c} + ac + \overline{y}.$$

By (13') and (13°), the negative of $\overline{a}\overline{b}\overline{c}$ is a+b+c. Similarly the negative of $\overline{x}\ \overline{y}$ is x+y and so the negative of (1) is

$$(\overline{\overline{a}\overline{b}\overline{c}} + abc + \overline{x}\ )(\overline{\overline{a}\overline{b}\overline{c}} + ac + \overline{y})(a + b + c)(x + y)$$

as predicted by Ladd-Franklin. Multiplying we get:

$$(\overline{\overline{a}\overline{b}\overline{c}} + \overline{\overline{a}\overline{b}\overline{c}}y + abc + abc\,\overline{y} + \overline{\overline{a}\overline{b}\overline{c}}\ \overline{x} + ac\,\overline{x} + \overline{x}\ \overline{y})(ax+ay+bx+by+cx+cy).$$

Using bb=b and $\overline{\overline{b}}$=0 and a+ab=a (Schröder's Law of Absorption), this becomes

$$(\overline{a}\ \overline{b}\ \overline{c}+abc+ac\,\overline{x} + \overline{x}\ \overline{y}\ )(ax + ay + bx + by + cx + cy).$$

Multiplying out we have,

$$0+0+0+0+0+0+ abcx+abcy+abcx+abcy+abcx+abcy+0+$$
$$ac\overline{x}y+0+ abc\overline{x}y + 0+ac\overline{x}y = abcx + abcy + ac\overline{x}y$$

or

No one who has not paid his subscription can be a member of all three sections and deliver a lecture or perform an experiment, or of a and c and perform an experiment without lecturing.

This however does not agree with Christine Ladd-Franklin's own conclusion of abcx + ac $\overline{x}y$ although it is very close. It is possible that Ladd-Franklin simply overlooked the middle term of abcy when stating her conclusion.

5.3.3. Mitchell's "On a New Algebra of Logic" of 1883. According to Dipert[47] Mitchell who was born in Ohio in 1851, had a short and tragic life. He grew up on a farm and was the eldest of eight children. It was presumably only the large number of younger siblings that freed him from duties on the farm. He became Principal of Marietta High School in Ohio and then spent three years at the Johns Hopkins University studying logic with Peirce and mathematics with J. J. Sylvester. Sylvester spoke highly of Mitchell:

> ...I would have been very glad, not to say proud, to have been myself the author of them

namely two papers on number theory published in the *American Journal of Mathematics* [12]. He received his doctorate in mathematics in 1882.

Although he was offered a Tyndall fellowship for study in England (apparently upon the recommendation of Sylvester), he turned it down and instead returned to become Professor of Astronomy and Mathematics in Marietta College, his undergraduate Alma Mater. He was later to regret this decision since he was not happy at Marietta citing overwork and frustration at the lack of time to devote to his mathematical and logical research. Although severe and unyielding, Mitchell was very modest. Slow and exact speech seems to have been his hallmark, and he also expected precision of expression in his students. However, he had patience - unlike Peirce - and would not proceed to a new topic until he was absolutely certain that the slowest student understood the point, apparently to the irritation of the other students. He is described as having an especially close and joyful relationship with his three sons, the oldest of whom was just five years old at the time of Mitchell's death. At the early age of 37 he died of pneumonia in 1889. A student of Peirce's along with Ladd-Franklin at the Johns Hopkins University, he introduced indices to algebraic logic in a way that Peirce recognized as the key to quantification. The law of inference enunciated in (Mitchell 1883) is that of 'take the logical product of the premises and erase the terms to be eliminated'. He emphasized relations in a way that Ladd-Franklin did not, stating 'every proposition in its ultimate analysis expresses a relation among class terms'. For the first time the limit of the language or notation being used is considered. Mitchell wrote:

> The universe of class terms, implied by every proposition or set of propositions, may be limited or unlimited.

He used roman letters for class terms, U for the universe of class terms, Greek letters for propositions and ∞ for the universe of relative terms (often called the universe of relation), or for the possible state of things, unlike Ladd-Franklin who does not clarify the distinction between propositions and class terms and Peirce, who used the same symbol ∞ for the universe of both class terms and for the universe of relation. However Mitchell used a similar technique to Ladd- Franklin in his use of truth-values when working with propositional terms, e.g. he used a universe of relation for his propositional terms which defines the possible state of things.

The key to Mitchell's method is the use of subscripts to indicate quantification. He uses the subscripts 1 or u. A universal proposition or one that may be conceived as concerning "all of . . . the universe of class terms" is denoted as F_1. The symbol 1 in fact refers to the quantity of times or cases in which the proposition holds following Boole in *Laws of Thought*. A particular proposition or one that concerns SOME OF U is denoted as F_u. F itself represents any linear combination of logical terms involving class terms and any sum of products of such terms.

The use of subscripts in itself was not particularly novel, given that Peirce had introduced superscripts and subscripts before in *Logic of Relatives*. Superscripts were used to indicate universal quantification in the form of his operation of involution for universal expressions, but subscripts were used in his operation of backwards involution to indicate converse operations rather than quantification. Mitchell introduced the concept of using the same notation, i.e., that of subscripts to indicate both universal and particular quantification thus opening up the way for the universal and existential quantifiers.

F_1 and F_u are negatives in that $\overline{\overline{F}}_1 = F_u$, where the longer line represents the negative of the proposition and the shorter line indicates the negative of the predicate F. F_1 and $\overline{F}_1$ are contraries of each other, i.e. $F_1\overline{F}_1 = 0$. It should be noted that Mitchell used 1 interchangeably with U the symbol of the universe of class terms.

The four traditional syllogistic forms E, I, A, O are represented:

E	$(\bar{a} + \bar{b})_l$	No a is b
I	$(ab)_u$	Some a is b
A	$(\bar{a} + b)_l$	All a is b
O	$(a\bar{b})_u$	Some a is not b

TABLE 5.1. Traditional Syllogistic Forms

Mitchell also developed a table of the standard sixteen propositions obtained by applying the two forms of the universal and particular to the sixteen possible sums of ab, $\bar{a}b$, $a\bar{b}$, and $\overline{ab}$.

Taking A and E and adding E' and A', (the two universal complements added by De Morgan to the classic two) we obtain as part of the table in its simplest form:

$(\bar{a} + b)_l$	A
$(\bar{a} + \bar{b})_l$	E
$(a + \bar{b})_l$	A'
$(a + b)_l$	E'

TABLE 5.2. Mitchell's Forms

Together with their negatives, they form the eight propositions of De Morgan. It can be seen that subscripts l and u are used to express quantification with F_l meaning ALL U IS F being used for universal propositions or all propositions F hold over the universe of class terms and F_u meaning SOME U IS F or SOME F IS TRUE where U is the universe of class terms. Mitchell gives the following inferences:

$$F_l \ G_l = (FG)_l$$
$$F_l \ G_u \prec (FG)_u$$
$$F_u \ G_u \prec \infty$$

The dual formulæ, e.g.,

$$F_u + G_u = (F + G)_u$$
$$F_u + G_l \prec (F+G)_u$$
$$F_l + G_l \prec (F+G)_l$$

are also given. The symbol ∞ is the universe of propositions, e.g., ALL TIME and u is assumed to be greater than zero and less than 1 or the Universe U. 1 is often used by Mitchell as a convenient shorthand for U.

Dipert considers that it is precisely Mitchell's use of different forms of possible universes that is one of the most important of Mitchell's contributions to logic in his short life. He writes:[14]

> But Mitchell has constructed a system of notation in which reference to more than one universe of discourse is possible, and it is here that Mitchell's greatest contribution lies. These multiple universes are described as the "dimensions" of the expression.

Dipert also notes that it was this distinction between the universe of class terms and the universe of propositions, i.e., ALL TIMES that probably motivated Schröder's distinctions among the universes of 2-place, 3-place, ... relations. One of major shortcomings of Boolean algebraic logic was its difficulty in expressing mixed propositional and categorical statements such as IF ALL SWANS ARE WHITE THEN NO SWANS ARE BLACK. By using different (limited) universes of discourse for objects SWANS signified by U and times for which propositions are true, signified by ∞, Mitchell had solved one of the major inadequacies of most Boolean theories.

5.3.4. Mitchell and the Symbols of Quantification Π and Σ. The symbols Π and Σ were used in [128] to denote a product and sum respectively of any logical polynomials of class terms. However their purpose was to describe linear combinations of logical terms in their most general form, e.g.,

$$\Pi(F_u + \Sigma G_1) \quad \text{or} \quad \Sigma(F_1 \Pi G_u),$$

not as quantifiers. Quantification was reserved for his subscripts 1 or u. Peirce himself had developed this form of notation in *Logic of Relatives*, as generalized disjunction and as generalized conjunction.

Particular propositions were however linked with existence if not with the existential quantifier. Mitchell wrote:[15]

---

[14]See [**47**, p. 525].
[15]See [**128**, p. 84].

A particular proposition implies the existence of its subject, while a universal does not.

As far as I can ascertain, this concept is not new but was first seen in Peirce and in fact he cited Peirce in a brief footnote "Mr. Peirce and others." Compare this with Peirce's explanation in "Note B"[16] in the same volume:

> We write lb for LOVER OF A BENEFACTOR, and $l\dag b$ for LOVER OF EVERY-
> THING BUT BENEFACTORS.
> The former is called a particular combination, because it implies the existence of something ... The second combination is said to be universal, because it implies the non-existence of anything except what is either loved by its relate or a benefactor of its correlate.

Another innovation of Mitchell was identifying universal propositions with the U symbol in the following way: any product of particular propositions, i.e., any product of those terms with u as subscript gives the most general proposition since "there can be no inference when nothing is known about the relation of the two suffices." But the most general proposition, i.e., a linear combination of class terms of one of the above forms can be expressed as the sum of products of the eight propositions of De Morgan previously listed. The universal form F_1 being equivalent to the product of one or more of the propositions 2, 3, 4, 5 so that $F_1 = \Pi$ a where a is one of the four universals of De Morgan and any particular form being equivalent to one or more of the last four propositions of De Morgan, so that $\Gamma u = \Sigma b$. Therefore

$$\Pi\Gamma u = \Pi\Sigma b = \Sigma\Pi b$$

since multiplication is distributive over addition. Thus

$$\Sigma(F_1\Pi\Gamma u) = \Sigma(\Pi a\Sigma\Pi b) = \Sigma(\Pi a\Pi b).$$

So a general proposition can be expressed as a sum of products of the eight propositions of De Morgan, i.e., any general proposition can be reduced to the sum of products of the eight propositions of De Morgan.

Mitchell's main aim in his 1883 paper was to form a rigorous method to produce conclusions from given formulae. Dipert suggests[17] that:

---

[16]See [157, p. 189].
[17]See [47, p. 521].

His broader interest could be helpfully described as the characterization of "mechanical methods" in the algebra of logic

Hypothetical propositions of the form "IF a THEN b" are represented by the disjunctive proposition $\bar{a}+b$. So IF A IS BC, THEN CD IS E is represented by

$$\overline{(a + bc)}_1 + (cd + e)_1$$

which Mitchell claims gives

$$(ab + ac)_u + (cd + e)_1.$$

I have analyzed the reasoning behind this in the following process: take the negative of the first term by means of changing + to x and taking the negatives of the individual propositions we have

$$(a(b + c))_u + (cd + e)_1,$$

which gives

$$(ab + ac)_u + (cd + e)_1.$$

On page 82 the solution to one of the problems in Boole's *Laws of Thought*, page 146 is given. Mitchell used the method of elimination which is the process of eliminating terms where by doing so no existing term vanishes, i.e., from a+ bcd then b, c, d, bc, bd or cd can be eliminated, but not a or bcd.

The premises are

(1) $\qquad (x+z+vy\ \bar{w}+vw\ \bar{y})_1$

(2) $\qquad (v + \bar{x} + \bar{w}+yz+ \bar{y}\ \bar{z})_1$

(3) $\qquad (\bar{x}+ \bar{v}\bar{y}+w\bar{z}+ \bar{w}z)_1\ (xy + vx + wz +\bar{w}\ \bar{z})_1$

Mitchell's main method of solution is to multiply the premises together and then add to the result x and y or alternatively eliminate x and y from the result and finally simplify the resultant expression. Notice that he works with expressions and not equations;

this is because equality is inherent in the subscript notation, e.g., $\overline{(a+b)}_1$ represents the Boolean equation $A(1 - B) = 0$ or ALL A IS B.

Let us consider more closely Mitchell's solution. By expanding (3) we get:

$$wxz + \bar{w}\bar{x}\bar{z} + \bar{v}w\bar{y}z + \bar{v}\bar{w}\bar{y}\bar{z} + wxy\bar{z} + vwx\bar{z} + \bar{w}xyz + v\bar{w}xz.$$

Multiplying by (2) and eliminating repetitions we get

$$vw\bar{x}z + v\bar{w}\bar{x}\bar{z} + vwxy\bar{z} + vwx\bar{z} + v\bar{w}xz + v\bar{w}xyz + v\bar{w}xz + w\bar{x}z +$$
$$\bar{w}\bar{x}\bar{z} + \bar{v}w\bar{x}\bar{y}\bar{z} + \bar{v}\bar{w}\bar{x}\bar{y}\bar{z} + \bar{v}\bar{w}\bar{y}\bar{z} + \bar{w}xyz + w\bar{x}yz + \bar{w}\bar{x}\bar{y}\bar{z} + vwx\bar{y}\bar{z}.$$

Multiplying by (1) and eliminating repetitions we get

$$vwxy\bar{z} + vwx\bar{z} + v\bar{w}xyz + v\bar{w}xz + \bar{v}\bar{w}x\bar{y}\bar{z} + \bar{w}xyz + vwx\bar{y}\bar{z} +$$
$$vw\bar{x}z + w\bar{x}z + \bar{v}w\bar{x}\bar{y}z + w\bar{x}yz+v\bar{w}\bar{x}y\bar{z} + v\bar{w}\bar{x}y\bar{z} + vwx\bar{y}\bar{z} + vwx\bar{y}\bar{z}.$$

Casting out v using the method of elimination and deleting repetitions we get

$$wxy\bar{z}+ wx\ \bar{z}+ \bar{w}xyz + \bar{w}xz +\bar{v}\ \bar{w}x\ \bar{y}\bar{z} +$$
$$wx\ \bar{y}\bar{z}+w\bar{x}z + \bar{v}w\bar{x}\bar{y}z + \bar{w}xyz+ \bar{w}\bar{x}y\bar{z}+w\bar{x}\bar{y}z.$$

Eliminating $\bar{v}$ from the expression and eliminating repetitions we get

$$wx\bar{z} + \bar{w}xz + \bar{w}x\bar{y}\bar{z} + w\bar{x}z + \bar{w}\bar{x}y\bar{z}.$$

However Mitchell obtains

$$wx\bar{z} + w\bar{x}z + \bar{w}x\bar{y} + \bar{w}xz + \bar{w}\bar{x}y\bar{z}.$$

This is more likely to be a slip than a typographical error since Mitchell uses the result in the explanation that follows. However according to his method the extra coefficient of $\bar{z}$ can be eliminated as long as any term in the linear combination does not equal zero so that his result is also valid. As he states quite clearly later on page 87 of his paper:

So in regard to elimination, any set of terms can be eliminated by neglect, provided no aggregant term is thereby destroyed.[18]

The main drawbacks to Mitchell's subscript notation for quantification have been pointed out by Dipert; namely that it is not possible to express unambiguously repeat occurrences of variables, or even priority of one quantifier over another, which is done in modern notation by the order of the quantifier expressions. Peirce was to solve this problem and furthermore introduce his own form of quantifier symbols in an article "Note B" in the same volume *Studies In Logic* as Mitchell's paper, suggesting that he was closely involved with the preparation of his student's work prior to publication. Since he always gave Mitchell credit for introducing a notation for quantification, it is likely that Peirce developed his own paper following Mitchells lead.

5.3.5. Peirce's "Note B: The Logic of Relatives" of 1883. This paper published together with the work of Ladd-Franklin and Mitchell in [**120**] developed quantifica- tion further and introduces a coefficient for his logical terms that implies existence in a similar way to that of the Boolean truth values: 1 for existence, 0 for non-existence. As early as 1870 in *Logic of Relatives*, Peirce had attempted some form of quantification in his operation of involution; in particular universal quantification, e.g., l^W represents the class of the LOVERS OF EVERY WOMAN. (Merrill 1997) has also noted that universal quantification to some extent had been inherited from Boole in the equational form h(1-a)=0 meaning ALL HORSES ARE ANIMALS and could also be expressed using Peirce's own copula of illation h≺a, although it must be remembered that Boole had no quan- tifiers. For existential quantification, Boole used the partial class symbol v meaning SOME.[19] However his equational representation led to error, as Peirce pointed out that using this form v,h = v,b meaning SOME HORSES ARE BLACK, and then negating this, SOME Xs ARE NOT Ys could be obtained from SOME Ys ARE NOT Xs.

Merrill points out that Peirce overcame this problem in two ways. First, by using the sign of inequality, h<b, to mean h≺b but it is not true that b≺h, so that SOME HORSES

---

[18]Mitchell also used a subscript e to express quantification This is an inclusive symbol meaning either u or 1. Similarly in nature to Boole's coefficient of quantity v, the quantifier e can be either 1 or u where u means greater than 0 but less than 1. Mitchell uses the following law to combine propositions: "(1) The conclusion from the product of two premises is the product of the predicates of the premises affected by a suffix equal to the product of the suffices of the premises, i.e., F_e $G_{e'}$ < $(FG)_{ee'}$. For the dual rule (2), the word *product* is replaced by the word *sum*.

[19]See Chapter 4.

ARE BLACK is represented by h,b > 0. Secondly he used his operation of involution, $0^{h,b} = 0$, to represent SOME HORSES ARE BLACK or that the class h,b has members since $0^{h,b}$ represents the class of all things in the null relation to every member of h,b. Here the first 0 represents the null relative and the second 0 the null class; 0 being the zero relative term such that $x + 0 = x$ and $x, 0 = 0$, vanishing when x exists, and not vanishing when x does not exist.

A very important area had yet to be addressed. To deal explicitly with the issue of mixed quantification, i.e., universal and particular quantification, Peirce followed the lead of his student Mitchell in representing the existential and universal quantifiers as operators on his logical terms but not in the form of attached subscripts but rather by using his symbols for infinite sums and products. Another radical shift in "Note B" is the change in emphasis from relative terms and classes to relations and ordered pairs. For the first time he defined a dual relative as determining an ordered pair of objects, e.g., A:B. It is to be understood however that at this stage Peirce always emphasized the first letter of the pair, identifying the relative with A rather than the implied relation LOVES. A general relative is defined as a linear combination or, as Peirce writes, *logical aggregate* of a number of such individual relatives.

If l denotes LOVER then
$$l = \Sigma_i \Sigma_j (l)_{ij} (I : J)$$
where the subscripts i and j signify that this sum is to be taken over all pairs of objects in the universe and $(l)_{ij}$ is 1 in the case that I is a LOVER of J and 0 otherwise. The negative of a relative is defined as its complement and is signified by drawing a straight line over the sign for the relative itself. The converse relative is written B:A. Peirce writes on page 188:

> Thus the converse of "lover" is "loved." The converse may be represented by drawing a curved line over the sign for the relative.

Negatives and converses are expressed very simply as
$$\overline{\overline{l}} = l \quad \text{and} \quad \overset{\smile\smile}{l} = l$$

as the formula "the negative of the converse of a relative is the converse of the negative of the relative." We also have the following formulæ:

$$(l \prec b) = (\bar{b} \prec \bar{l})$$

$$(l \prec b) = (\breve{l} \prec \breve{b})$$

The following equation then holds:

$$(l + b)_{ij} = (l)_{ij} + (b)_{ij}$$

where $(l)_{ij}$ is either 1 if I is a lover of J or 0 if not. It must be noted however that here 1 represents the universe so that we have $1+1=1$. We also have

$$(l, b)_{ij} = (l)_{ij} \times (b)_{ij}$$

where the comma signifies logical composition or Boole's multiplication called *non-relative* or *internal multiplication*. This, in words, is:

    I the class of lovers and benefactors of the class J consists of the class of lovers of J together with the class of benefactors of J.

The operations of composition and transaddition are called *relative multiplication* and *relative addition* respectively and are defined by the truth-values:

$$(lb)_{ij} = \Sigma_x (l)_{ix} (b)_{xj} \qquad (l \dagger b)_{ij} = \Pi_x \{(l)_{ix} + (b)_{xj}\}$$

In order to apply Peirce's logical algebra, four basic formulae are used which involve three pre-defined relatives. These are the relatives denoted by ∞ which is the universal relative or co-existent with, the identity relative 1, and the negative relative other than _____ or not signified by n. From these we obtain the following four key formulæ:

$$l, \bar{l} = 0 \qquad l\bar{l} \prec \mathbf{n}$$

and

$$l + \bar{l} = \infty \qquad l \prec l \dagger \bar{l}$$

For example, to eliminate s from the two propositions

$$l \prec l\bar{s} \qquad l \prec sb,$$

we relatively multiply them in such an order as to bring the two ss together:

$$l \prec l s \bar{s} b \prec l \mathbf{n} b$$

by applying the second of the above formulæ.

It is at this stage in "Note B," that Peirce makes the connection between quantification and the quantifier symbols Σ and Π. The Σ symbol had been introduced in *Logic of Relatives*, to represent finite sums. He also defined logical terms using an infinite sum of individuals. By 1880, Peirce had begun to define his relative terms both through the use of infinite sums and also products explicitly using Σ and Π as algebraic notation to represent such infinite sums and products.

Peirce makes this connection through the numerical coefficients that were introduced at the very beginning of the paper as truth-values or Boolean values and not referred to since. Of the numerical coefficients he states:[20]

> Any proposition whatever is equivalent to saying that some of the sums and products of such numerical coefficients is greater than zero.

Thus,

$$\Sigma_i \Sigma_j l_{ij} > 0$$

means SOMETHING IS A LOVER OF SOMETHING;

$$\Pi_i \Sigma_j l_{ij} > 0$$

means EVERYTHING IS A LOVER OF SOMETHING. Peirce then takes the major step of omitting the inequality symbol and final zero to associate universality with Π and existence or particular propositions with Σ so that these are no longer symbols for infinite products or sums but also quantifier symbols. In this way

$$\Pi_i \Sigma_j (l)_{ij}\, b_{ij}$$

means EVERYTHING IS AT ONCE A LOVER AND A BENEFACTOR OF SOMETHING.[21]

$$\Pi_i \Sigma_j (l)_{ij}\, b_{ji}$$

means EVERYTHING IS A LOVER OF A BENEFACTOR OF ITSELF and

$$\Sigma_i \Sigma_k \Pi_j (l)_{ij} + (b)_{jk}$$

means SOMETHING IS A LOVER OF EVERYTHING EXCEPT BENEFACTORS OF SOMETHING.

---

[20]See [**157**, p. 200].

[21]I disagree with Martin in his paper "Individuality and Quantification, Peirce's Logic of Relations and Other Studies" [**121**, p. 23], that the variables $(l)_{ij}$ are true logical terms representing classes but rather they are Boolean logical coefficients having values 1 or 0. Peirce may well have been thinking of the former case but he was careful in his 1883 paper to use the logical coefficient format.

Peirce went on to present the rules for mixed quantification, treating the quantification symbols as operators, e.g.,

$$\Sigma_i \Pi_i \prec \Pi_i \Sigma_i$$

We also have

$$\{\Pi_i \varphi(i)\}\{\Pi_j \psi(j)\} = \Pi_i \{\varphi(i) \cdot \psi(i)\}$$
$$\Pi_i \varphi(i)\}\{\Sigma_j \psi(j)\} \prec \Pi_i \{\varphi(i) \cdot \psi(i)\}$$

where i and j are individuals and φ and ψ are relatives. The expressions $\varphi(i)$ and $\psi(j)$ have not been previously defined but these probably refer to logical expressions of relative terms using only the commutative operations and the operations inverse to them where the variable i denotes the individuals in I, the domain of the relation as previously defined in *Logic of Relatives*, and with presumably logical coefficients attached.

5.4. Quantification

5.4.1. The Origins of the Quantifiers. It can be seen that Peirce makes extensive use of the symbols Σ and Π to express relative terms as the sum of individuals or alternatively the negative relative term is expressed as the product of negative individuals. The origins of this lay in his 1870 *Logic of Relatives* paper where an absolute term is defined as an aggregate of some of the individual things in the universe. He progressed from this to defining a relative term in a similar way, e g., in his paper "On the Logic of Relatives", a dual relative term, such as *l*, LOVER OF _____, is defined as an aggregate of pairs A:B.

In fact, prior to publication of "Note B," in an unpublished letter from Peirce to Mitchell dated 21 December 1882, Peirce expressed quantification by using the sum and product symbols for the first time, but did not yet dispense with the inequality, so this can be seen as an earlier stage in the process of identifying the existential quantifier with Σ and the universal quantifier with Π. For example,

$$\Sigma_x \Sigma_y b_{xy} l_{xy} > 0$$

is to be interpreted as SOMETHING IS BOTH A BENEFACTOR AND LOVER OF SOMETHING.

$$\Sigma_x \Sigma_y b_{xy} l_{yx} > 0$$

means SOMETHING IS A BENEFACTOR OF A LOVER OF ITSELF while

$$\Sigma_x l_{xx} > 0$$

means SOMETHING IS A LOVER OF ITSELF.

Similarly,

$$\Pi_x \Pi_y (l_{xy} + b_{xy}) > 0$$

means that EVERYTHING IS EITHER A LOVER OR A BENEFACTOR OF EVERYTHING and

$$\Pi_x l_{xx} > 0$$

EVERYTHING IS A LOVER OF ITSELF. It is interesting to note that Peirce does not use brackets to denote the Boolean coefficients, but they are implicitly implied through the use of the inequality symbol which indicates that some numerical value is attached to the term.

The quantifiers are used in the same way as they were used before in [120] but by dispensing with the inequality > and zero Peirce even more closely identifies his sum and product symbols as quantifier symbols or operators. In his subsequent paper "On the Algebra of Logic, A Contribution to the Philosophy of Notation" [158], he fulsomely credits Mitchell with the discovery of a notation for expressing quantification in logic. However, it was Peirce himself who made the step of then identifying such quantification with the Σ and Π symbols used for repeated sums (disjunction) and repeated products (conjunction) respectively. In fact as Zeman has shown by 1885, Peirce had in "On the Algebra of Logic, A Contribution to the Philosophy of Notation" a complete quantification theory with identity together with a system for expressing mixed quantification.[22]

These papers reaffirm novel ideas and notations already discussed in Section 1 of this chapter such as truth-values. Peirce attempted to interpret traditional syllogistic propositions and solve logical problems but had to admit "I shall not be able to perfect the algebra sufficiently to give facile methods of reaching logical conclusions."

5.4.2. Peirce's Development of the Quantifier. In tracing the development of the quantifier in Peirce's work, we may ask a number of questions. What are the quantifiers? Are Σ and Π symbols for infinite sums or products respectively? Is Peirce

---

[22]See [206, p. 7].

aware of the inherent difficulties of working with infinite classes or even an infinite language? When did the Σ and Π notation become identified with quantification?

Peirce first defined his logical terms, in particular his relative terms, by using infinite sums in *Logic of Relatives*. The Σ symbol was not used to represent such infinite sums at this stage, although it was used in *Logic of Relatives* to represent the sums of the binomial theorem which Peirce used as an algebraic analogy to provide the basis for his theory of logical differentiation. In "On the Algebra of Logic," Peirce extended his use of infinite sums to define his negative relative terms as an infinite product of class complements. The Σ and Π symbols were both used and subscripts indicated individuals, e.g., $l = \Sigma(L_i : M_j)$ where L_i and M_j are individuals, and similarly for products $l = \Pi(L_i : M_j)$. This method of defining relative terms as an infinite sum of individuals or an infinite product of complements was described by Peirce as the method of limits [**153**].

Mitchell's paper contained the next advance in quantification theory. He used subscripts to denote quantification, in that the subscript 1 indicated universal propositions:

$$(\bar{A} + B)_1$$

for ALL A IS B and the subscript u indicated particular propositions:

$$(AB)_u$$

for SOME A IS B.[23] This is similar to Peirce's own use of superscripts to indicate universality which first appeared in [**149**], in his operation of involution: l^s denoting WHATEVER IS THE LOVER OF EVERY SERVANT OF _____.

As far as the symbols Σ and Π are concerned Mitchell did not identify them as symbols of quantification, reserving these for the subscripts 1 and u, as seen earlier. Mitchell did note that particular propositions were related to existence, a concept that he had obtained from Peirce.[24] Furthermore he combined the Σ and Π notations to obtain linear combinations of logical terms in their most general form. This came from his interest in expressing a general proposition in terms of a product of De Morgan's syllogistic propositions ALL A IS B, etc. Mitchell's achievement was to link De Morgan's traditional syllogistic forms with a workable method of obtaining conclusions

---

[23]See [**128**].
[24]See Section 2.2.1 above.

using a subscript notation for quantification, both of subject and predicate. He broadened the traditional copula ıs to cover Peirce's inclusion copula, $\prec$. However because he did not use relative terms he had to work within a propositional framework.

It was left to Peirce to link quantification with sum and products and in particular with the symbols representing sums and products. He always gave Mitchell full credit[25] for introducing a notation for expressing quantification. As editor of *Studies in Logic* he had obviously seen Mitchell's paper before publication and although in his "Note B," the concept of quantification and method of working with the algebraic logic are essentially those of Mitchell, his own notation is very different.

His first attempt in his 1882 letter combined the Σ and Π symbols with an inequality symbol and 0 to imply existence, e.g.,

$$\Sigma_y \Pi_x (l_{xy} + b_{xy}) > 0$$

or THERE IS SOMETHING OF WHICH EVERYTHING IS EITHER LOVER OR BENEFACTOR. This is repeated in his "Note B" in *Studies in Logic*, but here Peirce uses a Boolean coefficient $(l)_{ij}$ to indicate existence of a relative term l. $(l)_{ij}$ has the value 1 if the proposition ı LOVES ȷ holds, where i and j are individuals and the value 0 if the proposition is false, so that

$$l = \Sigma_i \Sigma_j (l)_{ij} (I : J).$$

To say that SUCH A LOVER ı OF ȷ EXISTS is equivalent to

$$l = \Sigma_i \Sigma_j (l)_{ij} > 0.$$

This represents a combination of concepts that Peirce had been developing for a decade namely modal logic as in truth-values, and a means of expressing propositions by using relative terms represented by an infinite sum of individuals. It is strange that this 1883 paper represents a step backwards from his 1882 letter to Mitchell in that quantification is only expressed in terms of these numerical coefficients $(l)_{ij}$ or it could be the case that in the Mitchell letter the Boolean coefficients were implicit. I do not however think this is likely.

---

[25]In [158] Peirce wrote "All attempts to introduce this distinction [quantification] were more or less complete failures until Mr Mitchell showed how it was to be effected."

In any case, Peirce soon realized that it was <u>sufficient</u> to represent a proposition using only the numerical coefficient $(l)_{ij}$ and the quantifying symbol, e.g.,

$$\Sigma_j \Pi_i \prec \Pi_i \Sigma_j$$

means that EVERYTHING IS AT ONCE A LOVER AND BENEFACTOR OF SOMETHING. He also began to treat the quantification symbols as operators obeying certain rules such as

$$\Sigma_j \Pi_i \prec \Pi_i \Sigma_j$$

By 1885, Peirce was expressing quantification without recourse to numerical coefficients. Propositions were now composed of a Boolean expression referring to an individual and a quantifying part specifying the individual.

5.4.3. Limited and Unlimited Universes. In his earlier papers, Peirce had avoided the pitfalls of an all-embracing universe. by using De Morgan's concept of the universe of discourse where the universe is limited to those individuals or qualities under discussion. Peirce demonstrated this in [**158**]: any $(\bar{k} + h)$ means that any individual in the (limited) universe is either, not a king or is happy. There is a sense that Peirce felt that the move from a universe of individuals to a universe of relative terms involved a transition to a two dimensional universe. Mitchell himself used a second universe - that of time - and indicated two dimensional propositions by means of subscripts, e.g., $(bs)_{UV}$ means that SOME OF THE BROWNS SPENT PART OF THE SUMMER AT A VILLAGE where U is the universe comprising inhabitants of a certain village and V is the universe of time.

For the first time in this 1885 paper, he moved away from his central concept begun in *Logic of Relatives*, of relative terms where *l*, the relative term, was either represented by the class of WHATEVER IS A LOVER OF _____ or by a linear combination of dual pairs (I:J), to one where *l* now represents the relation LOVES. This change is most clearly stated in [**160**] as the change from classes to relations or as Peirce wrote:

> The best treatment of the logic of relatives, as I contend, will dispense altogether with class names and only use . . . verbs.[26]

This was probably effected because propositions could now be simplified to variable and quantifying symbol rather than also involve the Boolean coefficient $(l)_{ij}$ which took the value 1 if such a lover existed and 0 otherwise, as used in [**157**]:

---

[26]See [**84**, p. 290].

If x is a simple relation, $\Pi_i\Pi_j x_{ij}$ means that every i is in this relation to every j, $\Sigma_i\Pi_j x_{ij}$ that some one i is in this relation to every j ...

So, although Peirce had arrived at this notation, through expressing particular propositions as infinite sums of relative terms with their associated Boolean coefficients; by dispensing with these coefficients for the sake of simplicity he faced the problem of identifying the quantifying symbols with infinite sums and products of relative terms which are not bound by the universe of discourse. By using analogy, Peirce sought to deny that his quantifying symbols, Σ and Π, which he now referred to as *Quantifiers*, were true sums and products. He stated:

> ... in order to render the notation as iconical as possible we may use Σ for some, suggesting a sum, and Π for all, suggesting a product.

He later continued:

> It is to be remarked that $\Sigma_i x_i$ and $\Pi_i x_i$ are only similar to a sum and a product; they are not strictly of that nature, because the individuals of the universe may be innumerable.

As Beatty writes:[27]

> The quantifiers Σ and Π only suggest a sum and product; they may operate upon an infinite number of individuals but a sum and product may not.

Peirce stated clearly that in treating classes collectively he was aware that he was neither speaking of a single individual nor of a small number of individuals but of a whole class, perhaps an infinity of individuals. This of course, suggested a relative term with an indefinite series of indices. Two methods were then suggested by Peirce to deal with these unlimited universes. One method involved using indices of indices as in $\Sigma_a i_a$ is where we are to take any collection whatever of *i*s and then any individual of that collection. The other method is to restrict certain classes to finite collections. Peirce stated that such a restriction was sometimes necessary otherwise the following paradox would hold:

---

[27]See [**8**, p. 236].

Some odd number is prime;
Every odd number has its square, which is neither prime nor even;
Hence, some number is neither odd nor even.

Neither of these methods however was subsequently used by Peirce in any later work.

It has been noted previously that Mitchell's use of distinct universes of discourse differentiated between universes of objects, times, and qualities. Peirce greatly appreciated these multidimensional logical universes and praised Mitchell's work in this area as one of his most important contributions to exact logic. However Peirce used this concept in a different way. For him logical dimension represented the concept of BEING AN ELEMENT OF in Cantorian set theory and even in *Logic of Relatives*, he certainly was able to extend individuals in one dimension so that they could exist in different universes - in particular his extension of individual members, e.g., MAN to relative terms, e. g., MAN THAT IS _____.[28]

5.4.4. Quantifier Order. Although Peirce never attempted a general axiomatic development of quantification theory, as early as [157], he already had produced the following general rules:

a) $\Sigma_i \Pi_j < \Pi_j \Sigma_i$

b) $\{\Pi_i \phi(i)\} \{\Pi_j \psi(j)\} = \Pi_i \{\phi(i) \cdot \psi(i)\}$

c) $\{\Pi_i \phi(i)\} \{\Sigma_j \psi(j)\} < \Sigma_i \{\phi(i) \cdot \psi(i)\}$

In [158, p. 231], he described a seven step method of uniting a given set of premises and eliminating certain letters from them. Let us take as an example the following two steps:

Step 1 The quantifiers can be brought to the left hand side, e.g.,

$$\Pi_i x_i \cdot \Pi_j x_j = \Pi_i \Pi_j x_i x_j$$

---

[28]Martin notes that Peirce had recognized implicitly a narrower and a wider sense of universe of discourse in *Logic of Relatives* ranging over collections of objects and qualities [121, p. 13].

Step 2 As far as possible the Σs should be carried to the left of the Πs and subscripts may be rearranged in alphabetical order, e g.,

$$\Pi_j \Sigma_i x_i y_j = \Sigma_i \Pi_j x_i y_j$$
$$\Pi_j \Pi_i x_{ij} = \Pi_i \Pi_j x_{ij}$$
$$\Pi_j \Sigma_i x_i j_j = \Sigma_i \Pi_j x_i y_j$$

We have however,

$$\Pi_j \Sigma_i x_{ij} \succ \Sigma_i \Pi_j x_{ij}$$

when the is and js are 'not separated.'

The fifth step is described by Peirce thus (CP3,232):[29]

> The next step consists in multiplying the whole Boolian part, by the modification of itself produced by substituting for the index of any Π any other index standing to the left of it in the Quantifier. Thus, for $\Sigma_i \Pi_j l_{ij}$ if we can write $\Sigma_i \Pi_j l_{ij} l_{ii}$.

The sixth step consists of

> ...the re-manipulation of the Boolian part, consisting, first, in adding to any part any term we like; second, in dropping from any part any factor we like ...

so that $\Sigma_i \Pi_j l_{ij} l_{ii}$ becomes $\Sigma_i \Pi_j l_{ii}$. Also $x\bar{x} = \mathbf{f}$ and $x + \bar{x} = \mathbf{v}$. The seventh step is to eliminate any Quantifier whose index no longer appears in the Boolian, e.g., $\Sigma_i \Pi_j l_{ii}$ becomes $\Sigma_i l_{ii}$.

Let us now consider a problem-solving example as given in [**158**], CP3.397):

From the premises [sic] $\Sigma_i a_i b_i$; and $\Pi_j(\bar{b}_j + c_j)$, eliminate b.

The method used can be identified in the following steps. Using Steps 1 and 2 to multiply these expressions together we get $\Sigma_i \Pi_j a_i b_i(\bar{b}_j + c_j)$. Multiplying out by b_i we obtain $\Sigma_i \Pi_j a_i(b_i \bar{b}_j + b_i c_j)$. The second occurrence of b_i can be eliminated by Step 6.

---

[29]See [**84**, p. 232].

Next consider $\Sigma_i\Pi_j b_i\bar{b}_j$. Using Step 5 we can write $\Sigma_i\Pi_j b_i\bar{b}_i$, and Step 6 allows us to dispense with Π_j so that we obtain $\Sigma_i b_i\bar{b}_i$ which can be eliminated to get the conclusion $\Sigma_i a_i c_j$.

Peirce does not provide an English interpretation of this problem but I would suggest (using other instances where such a 'translation' is provided in the same paper) that we have as the premises: SOME MAN i IS BOTH AN ANGEL AND A BENEFACTOR for $\Sigma_i a_i b_i$ and EVERY MAN j IS EITHER NOT A BENEFACTOR OR IS A CHIMERA for $\Pi_j(\bar{b}_j+c_j)$. The result is then to be interpreted as THERE IS SOME MAN i WHO IS BOTH AN ANGEL AND A CHIMERA. This follows because since MAN i does not fall into the first category of the second premise, i.e., he is a BENEFACTOR; this means he must be a CHIMERA.

It can be seen from the above rules that Peirce had a very clear idea of quantifier order and used these as part of his method of obtaining inferences within his quantification logic.

5.5. Summary

Over a period of twenty seven years, from 1870 to 1897, Peirce developed his algebraic logic from its first conception in *Logic of Relatives* to a sophisticated and encompassing logic in [**160**], where a method for producing logical inferences from a set of premises was outlined. Starting from three logical terms - individuals, relative terms and conjugative terms, Peirce worked with the logic of classes. Individuals were classes with one member, and in the case of relative and conjugative terms, these were identified with classes.

Merrill does not agree that these relative terms stand for classes but prefers to consider them as sets of ordered pairs. Although as we have seen, Peirce came to consider them in this way by 1885. Merrill cites Lewis's argument 'If $l \prec s$, then $l^w \prec s^w$' implies

IF ALL LOVERS ARE SERVANTS, THEN A LOVER OF EVERY WOMAN IS A SERVANT OF EVERY WOMAN.

He points out correctly in [**125**] that this does not hold. Lewis has misinterpreted s the relative term WHATEVER IS A SERVANT OF _____ with s SERVANT an individual servant which represents the class of all servants. Merrill also considers s to stand for the class of servants - he writes:

> For instance, ls would have to stand for the class of lovers of servants; yet this is not a function of the class of lovers and the class of servants.

It is true that Peirce often dropped the first correlate to obtain LOVER OF but this is only because this is understood to mean WHATEVER IS A LOVER OF _____. Also note the fact that Peirce also considered his individuals to represent classes, e.g., w which stands for A WOMAN could equally be represented by w, or the class WHATEVER IS A WOMAN THAT IS _____.

Although we disagree with Merrills tentative conclusion that Pierce was quite clear that his relative terms stood for relations could be argued with, it is however true that it was the Boolean legacy that led Peirce to embed their logic within compound class terms. It is also true that in his papers post 1885, this restriction was dropped. In fact this early emphasis on classes rather than on relations was later bitterly regretted by Peirce. He wrote in an uncharacteristically humble note for Peirce:

> I must, with pain and shame, confess that in my early days I showed myself so little alive to the decencies of science that I presumed to change the name of this branch of logic [the logic of relations], a name established by its author and my master, Augustus De Morgan, to "the logic of relatives." I consider it my duty to say that this thoughtless act is a bitter reflection to me now, so that young writers may be warned not to prepare for themselves similar sources of unhappiness.

However he added disingenuously:[30]

> I am the more sorry, because my designation has come into general use.

------

[30]See [**162**, p. 367].

In *Logic of Relatives* Peirce also defined the operations of addition (aggregation but counting common individuals only once), multiplication (logical - 'inclusive and' and relative - 'composition of relations'), and involution l^s - 'WHATEVER IS THE LOVER OF EVERY SERVANT OF _____. The copula was either identity, =, or illation, ≺, meaning inclusion or inference - 'therefore'. This copula of inclusion does involve implicit universal quantification of the form EVERY A IS B. In quantificational terms, it is the operation of involution, which is most interesting in that it is 'one of the ways of dealing with quantification even if it did not contain quantifiers' mentioned in [**125**, p. 158]. Here it is only the second relative term that is quantified universally. It should also be noted that conjugative terms (two place relatives) and composition of relative terms (which form conjugatives) imply existential quantification. This can be readily understood to mean that as in composition of relations, the range of the first relation must exist in order to become the domain of the second relation.

He described universal propositions in a Boolean way: 'h,(1-a)=0' or ALL HORSES ARE ANIMALS or using illation h≺a. For particular propositions he used h,b>0 to mean SOME HORSE IS BLACK where h≻b but it is not true that b≻h. This can also be expressed by using Peirce's quantifying operation of involution as $0^{h,b} = 0$, where the first 0 stands for the null relation and the second occurrence is the null class (or in other words members of h,b exist). Merrill shows that in thus maintaining Boole's equational program of expressing all propositions as equations, Peirce was provided with an initial reason for quantifying his logic of relatives.

In later papers in the intervening twenty years, Peirce added innovations such as truth-values, and a fourth operation *transaddition l†s*, WHATEVER IS A LOVER OF EVERY-THING BUT SERVANTS OF _____. If propositions were to be substituted for his class terms, in order to satisfy logical equations they were 'true' or 'valid' - 'If A is true then B is true'. He expressed relative terms as infinite sums of individuals and negative relative terms as infinite products of complements. [**96**] has pointed out that the aim of Peirce at this stage in his 1880 paper "On the Algebra of Logic" was to apply his algebraic logic to syllogistic propositions and principles. Houser also claims that by taking six of Peirce's axioms, four definitions and three rules then we have a complete base for the classical propositional calculus.

By 1885, using Mitchell's inclusion of quantification in the form of SOME and ALL with propositions, but rejecting Mitchell's subscript notation[31], Peirce used the symbols Σ and Π as the quantifiers SOME and ALL together with Boolean coefficients that took the value 1 if that relation existed and 0 otherwise. An inequality was then sufficient to indicate existence: $\Sigma_i \Pi_j (l)_{ij} (\text{I:J}) > 0$ means that THERE IS SOME LOVER I OF SOME INDIVIDUAL J. This is then redefined so that l_{ij} now indicates the proposition that i is a lover of j and we have $\Sigma_i \Pi_j (l)_{ij}$. By dispensing with the inequality, Peirce now has to take account of infinite sums and products. This he does by using Σ and Π as icons suggesting sums and products but in fact represent quantifying operators. The very naturalness of this progression from the algebraic logic of relatives to this quantificational logic of relatives has been noted by Merrill. Peirce may have come to identify Σ with the existential operator through the fact that if his sums of individuals have a non-zero result, this implies existence and therefore particular propositions may be expressed in this way. He had been aware of this in 1870, when he expressed classes as sums of individuals. Identifying Π with universal propositions may have been suggested by Mitchell in his 1883 paper where he expressed universal propositions with products of the standard De Morgan propositions. Mitchell used Σ and Π only as symbols for repeated (not infinite) sums and products to express general forms of propositions, i.e., linear combinations of logical terms. The fact that he limited his sums and products to De Morgan's eight propositions meant that he did not have the difficulties associated with an infinite language.

In contrast to this, Peirce begins with the use of Σ and Π to represent infinite sums and products - but only of 1s and 0s (his logical coefficients). When he asserts that the quantifiers are only similar to sums and products it was because he was aware that the logical product and sum are functions defined as operations on a finite class whereas the quantifiers have to range over universes with possibly an infinite number of members. He wrote in [**84**, p. 228]:

Thus $\Sigma_i x_i$, means that x is true of some one of the individuals denoted by i or $= x_i + x_j + x_k \ldots$

and later:

---

[31][**44**] suggests that Peirce later forgot precisely what Mitchell's proposals had been but went on praising his work very highly, especially after Mitchell's premature death in 1889.

they are not strictly of this nature, because the individuals of the universe may be innumerable'.

Peirce does not rule out an infinitary language. In [59, p. 186] he discusses indices of indices as in Π_{i_α}:

> The necessity of some kind of notation of this description in treating of classes collectively appears from this consideration; that in such discourse we are neither speaking of a single individual (as in the non-relative logic) nor of a small number of individuals considered each for itself, but of a whole class, perhaps an infinity of individuals. This suggests a relative term with an indefinite series of indices as $x_{ijkl...}$.

This use of second-order quantification where the quantification ranges over relations between the terms rather than the terms themselves is shown by Brady to have enabled Peirce to express algebraic notions such as one-to-one correspondence. She writes:[32]

> The variables representing relations occur as subscripts, as they did in the first order case. It is in second intensional logic that mathematical notions appear. For example, he defines one-to-one correspondence using second-order quantifiers ...In general we find definitions and computations freely passing into second-order in his work and higher-order quantifiers to be frequently used as a tool for expressing mathematical reasoning.

In [160] he redefined logical terms so that they were no longer associated with classes, but with relations, e.g., LOVES and their associated ordered pairs. The three orders of logical terms are those of 'hecceity' or individual (as opposed to idea), also called 'monad' or 'monadic relative', and dyad or 'dyadic relative'- an aggregation of pairs, culminating in 'polyads' - ordered n-tuples.

The method of using this logic which consists of quantifiers and a Boolean part is essentially the same as that of Mitchell, i.e., any factor of a logical terms may be eliminated as long as the number of elements in the linear combination remain the

---

[32]See [16, p. 188].

same. Logical expressions may be multiplied together and terms added or eliminated as necessary to obtain a conclusion. This quantificational logic of relatives used only the two operations supplied by the quantifiers, neither being a relational operation. In his rules for a method of simplification and elimination for his quantificational logic he lists the following:

A) The Σs and Πs are rearranged so that the Σs stay to the left as far as possible, e.g., $\Pi_j \Sigma_i x_i j_j = \Sigma_i \Pi_j x_i j_j$. Peirce writes:[33]

> There will often be room for skill in choosing the most suitable arrangement.

One interpretation of this is FOR EVERY j, THERE IS SOME i SUCH THAT i IS A SERVANT AND j IS A LOVER is equivalent to THERE IS SOME i SUCH THAT i IS A SERVANT AND EVERY j IS A LOVER. This should however be distinguished from $\Pi_j \Sigma_i x_{ij} \prec \Sigma_i \Pi_j x_{ij}$. as on page 224 when the is and js are "not separated" which has a different scope. One interpretation of the left hand side could be FOR EVERY j THERE IS SOME i SUCH THAT i LOVES j and similarly the right hand side could be interpreted as THERE IS SOME i SUCH THAT i LOVES EVERY j.

B) Σs and Πs in the Quantifier part whose indices no longer appear in the Boolean are eliminated.

The advantage of this quantificational logic is that it has greater powers of expression in that there are propositions which can now be expressed which could not be in the *Logic of Relatives* relative notation. One is the Korselt result which Löwenheim reported in 1915, that states that the proposition that there are at least four individuals cannot be expressed in relative logic even though it can be in quantificational logic. Another factor, suggested by Merrill, in the evolution of the theory of quantification could have been the need to discover a convenient way of handling plural relatives which was never satisfactorily handled in *Logic of Relatives*. In terms of convenience of expression of propositional logic, it also scores more highly than the relational algebraic logic. Quantificational logic also has a uniform format for expressing propositions unlike the variety of possible expressions of Peirce's relational logic.

---

[33]See [**84**, p. 231].

The introduction of Mitchell's subscript notation for quantification inspired Peirce to experiment with the logical properties of his own quantifiers as operators on propositional functions. He then added them as a new operation to his algebra of relations. His second order quantification involved quantifying over all relations on a domain. This feature was necessary in order to formulate second-order mathematical properties such as induction and the least upper bound axiom in number theory. In [16] Brady asserts that in this way Schröder following on from Peirce, handled part of the foundation of mathematics within his theory. Schröder was to go further than Peirce in his attempt to show that the algebra of relations was an adequate basis in which to develop mathematics. The next chapter will compare the logics of Peirce and Schröder and give an example of Schröder's problem-solving technique.

CHAPTER 6

Comparison of the Logics of Peirce and Schröder

6.1. Influences on the Logic of Schröder

Friedrich Wilhelm Karl Ernst Schröder was born on November 25th 1841 in Mannheim, in the northern part of the German state of Baden. He was the oldest son of Heinrich Georg Friedrich Schröder, the director of the Higher Public School there, and received his education at the Universities of Heidelberg and Konigsberg in Germany. He was the professor of mathematics at the Institutes of Technology at Darmstadt and Karlsruhe. As far as personality went, he was a gentle and even-tempered man with a great deal of self-control – a complete contrast to Peirce. As a boy he had a facility for languages, mathematics and physics. He was a lonely and overworked individual who worked late into the night on his logic on top of his normally extensive teaching duties. He loved all kinds of sports and his death followed a cold after he had taken an extended bicycle trip. He died in Karlsruhe on June 16th, 1902 after an illness of several days that was diagnosed simply as 'brain fever."[1]

The first work on algebraic logic published by Schröder was entitled *Der Operationskreis der Logikkalkls* (*The Range of Operation of the Logical Calculus*) in 1877. Covering 37 pages, it adopted Boole's algebra of logic, with improvements suggested by Peirce and Jevons. It is a mathematical expression of logic rather than mathematics viewed as a theory of logic and showed the influences of Boole and the German mathematicians Hermann and Robert Grassmann. His major work was *Vorlesungen uber die Algebra der Logik* (*Lectures on the Algebra of Logic*), published in three volumes from 1890 to 1905 (vol. 1 in 1890, vol. 2 part 1 in 1891, vol. 3 part 1 (and only)

---

[1] See [**46**, p. 127].

in 1895, vol. 2.2 in 1905 posthumously). This included the logic of classes, of statement connections and of relations together with an elaborate treatment of part-whole theory.[2]

Dipert comments:[3]

> The impression one might however get that the mighty *Vorlesungen* were lectures in a demanding multi-year course of study on logic is almost surely false. There is no evidence of his having used them in teaching other than occasionally, or that he had other than isolated students who could have followed the material ... Because of its size, its still untranslated German text, the relative unpopularity of the algebra of logic in the German-speaking world of the 1890's, and the now-dated nature of its content, I would myself guess that only a very small handful of individuals have ever read carefully the *Vorlesungen* in its entirety.

The early influences on the logic of Schröder came from Boole and especially the Grassmann brothers. Like Boole, Schröder was concerned with problem solving encompassing and extending syllogistic logic and the process of logical reasoning in the form of elimination of logical terms expressed in algebraic equations. Although using many Boolean principles and concepts including the indeterminate Boolean class v, Schröder developed the mathematical analogies further by investigating the exhaustive general solutions of logical expressions. In this respect he was criticized by Peirce for overemphasizing mathematical methods at the expense of the logical concerns. By 1879, Schröder was familiar with Peirce's logic of relatives and incorporated this in his later logical work, so that he was a true successor to Peirce. He was also influenced by Peirce's theory of quantification as detailed in the previous chapter. It is clear that his debt to Peirce was great, as he was able to progress from the non-relative logic of Boole to the algebraic logic of Peirce incorporating the theory of relations and on to a logic with quantifiers.

However he does not appear to have had much influence on Peirce apart from his use of duality which began to be seen in the work of Peirce and his followers from 1880

---

[2]This work will be referred to henceforth as *Vorlesungen* together with the appropriate volume number.
[3]See [**46**, pp. 122–123].

onwards (although De Morgan also used duality in his theory of relations). Grattan-Guinness has pointed out that Schröder was the first logician systematically to explore duality in algebraic logic. He writes:[4]

> Boole, Peirce and others had noted features of duality, such as the pairing of connectives and of special classes, but it was Schröder who first made it a prominent feature of his system. All through his *Vorlesungen* he presented assumption, definitions and theorems in dual pairs whenever possible, even to the extent of splitting his page down the middle into two columns.

We will now investigate in more detail the similarities and differences in the logics of Schröder and Peirce, concentrating firstly on the problem-solving techniques of Schröder using a Ladd-Franklin example given earlier.

6.2. Problem Solving Techniques in Schröder

Schröder formalized the Duality Rule which stated that in a given law the operations + and x may be interchanged like 0 and 1 to form new laws. Many instances of this Duality principle with its dual laws were previously noted by De Morgan, the Grassmanns and Peirce. As well as his systematic inquiry of the duality doctrine in algebraic logic which went much further than Boole or Peirce, he also built an axiomatic approach to the propositional calculus adjoining it to his part/whole calculus of classes and then developing after Peirce the theory of quantification. Schröder proposed the following systems of axioms (where the symbol $\langle$ indicates class inclusion, + indicates union of classes while . indicates class conjunction):

1) $x \langle x$
2) If $x \langle y$ and $y \langle z$, then $x \langle z$
3) $(x + y) \langle z$ iff $x \langle z$ and $y \langle z$
4) $x \langle (y.z)$ iff $x \langle z$ and $x \langle y$
5) $x(y + z) = x.y + x.z$
6) $x \langle 1$ where 1 denotes the universal class
7) $0 \langle x$ where 0 denotes the empty class

---

[4]See [**65**, p. 121].

8) $1 \nmid x + x'$ where x' is the complement of x relative to 1

9) $x.x' \nmid 0$

Influenced by Peirce, he produced a detailed exposition of the logic of relations. He used Boole's 'Law of Development' in order to formalise deductive logic, but followed Jevons and Peirce in using the inclusive model of disjunction. However he placed at the foundation of the class calculus, Peirce's copula of class inclusion, rather than the equality favored by Boole. Schröder's relation of 'Subsumption' or 'Einordnung' $\nmid$ is not the elementhood operation of set-theory. It amounts (in its context, i.e., that of the part/whole theory of classes) to the operation of class inclusion $\subseteq$, but in the propositional calculus along the lines suggested by Peirce, $A \nmid B$ signifies $A \succ B$ but it is not true that $B \succ A$ (non-symmetric).

Schröder used 'normal forms' of logical expressions, i.e., equivalent linear combinations. The multipliers in the products were represented either by letters or letters with the subscript $_|$ to represent complements. This symbol should not be mistaken with 1 which represents the universe. Continuing the mathematical analogy, he was also interested in finding the solutions of logical equations. He emphasized mathematical terms and constructions. According to his definition, a solution of a given logical equation of the form $ax + bx_| = 0$ for two distinct classes a and b, where $x_|$ is the complement of x. The complement is that value of x whose substitution in the equation gives the same expression as would be obtained by eliminating x from the equation (i. e. results in the equation $ab = 0$). He obtained as the solution

$$x = a_| b + u a_| b_|$$

where u is used in the same sense as v the Boolean indeterminate class meaning 'some, all or none'. However Schröder rules out the meaning 'all', as we shall see in the following analysis of his method:

Problem: to obtain

$$x = a_| b + u a_| b_|$$

as a solution of

$$ax + bx_| = 0$$

On page 458 of volume 1 of the *Vorlesungen* he writes that the equations

$$x) \quad x = a_| u + b u_| \text{ and } x_| = au + b_| u_|$$

are given as solutions of the equation $ax + bx_| = 0$, and indeed these equations do give $ab = 0$ when the expressions for x and $x_|$ are substituted. Equation $x)$ claimed Schröder, can be written as $\lambda) \ x = b + a_| u$ where u is non-empty, and we shall investigate the reason for this later. It is interesting to note that it is exactly this equation that Peirce obtained in "Harvard Lecture VI"[5] as a solution to $ax = b$ to define the operation of division b/a. However, for Peirce to have a meaningful operation of division, a necessary condition was that a should contain b so that ab is non-empty.

Substituting in x) the normal forms $a_|b + a_|b_|$ for $a_|$ and $ab + a_|b$ for b we have

$$\mu) \ x = (ab + a_|b)u_| + (a_|b + a_|b_|)u = a_|bu_| + (a_|b + a_|b_|)u$$

since $ab = 0$. Then μ) becomes

$$x = u_|a_|b + ua_|b + ua_|b_|.$$

This is not the corresponding expression $a_|b + ua_|b_|$ given by Schröder. How are we to account for this, as Schröder does not provide any explanation? The key is the concept of the class u or 'Parameter' which is such that ua gives 'some, all or none' of a. However, it is clear that u is not used in this Boolean sense as the meaning 'all' is ruled out. So that u is a proper subclass of a and the complement $u_|$ can never be 0.

Since we have $ab = 0$, it follows that $bu = 0$ and $bu_| = b$ as in equation λ). In this way $a_|bu_|$ can be written as $a_|b$ and $ua_|b$ is zero, so giving the required result

$$\nu) \ x = a_|b + a_|b_|u.$$

This expression $a_|b + ua_|b_|$ for the solution x is the same result as that obtained by solving the equation $ax + b(1 - x) = 0$ by Boole's method.

Let us analyze Schröder's algebraic methods in more detail. In order to solve problems Schröder transformed his subsumption copula to equality in order to maximize the analogy with algebraic equations. He used two main techniques.

A) He replaced $a \nleqslant b$ with the equation $ab_| = 0$ where $b_|$ is the complement of b.
B) He eliminated a variable, e.g., a by expressing the resulting equation in normal form, i.e., $am + a_|n = 0$ which gives the result $mn = 0$. This follows because if $am = 0$ and $a_|n = 0$ then $mn = 0$ since n is a subclass of a.

On page 528 of volume 1 of *Vorlesungen*, we have a problem given by Venn. This happens to be exactly the same problem that we have analyzed previously with the solution given by Ladd-Franklin in Chapter 5, page 217 above. I will give the English translation:

---

[5]See Chapter 3 above.

The members of a board were all of them either bond-holders or share-holders, but no member was bond-holder and share-holder at once; and the bond-holders, as it happened, were all on the board. What is the relation between bond-holders and share-holders? Solution. Translation of the data into symbols yields:

$$a \nmid bc_| + b_|c, \quad b \nmid a.$$

Schröder then supplied two equations before he reached the conclusion $bc = 0$.

(1) $$a(bc_| + b_|c)_| + ba_| = 0$$

(2) $$b(bc + b_|c_|) = 0$$

and finally on expansion

$$bc = 0.$$

Let us look more closely at how equation (1) is obtained since further calculations are not provided. Applying the first of Schröder's techniques, i.e., the principle (A) that $a \nmid b$ implies $ab_| = 0$, the two subsumptions are combined into the single equation (1).

Looking at $(bc_| + b_|c)_|$ this becomes on taking the complement $(b_| + c)(b + c_|)$, and on expansion we get $b_|b + b_|c_| + bc + cc_|$ which equals $b_|c_| + bc$. This explains how Schröder obtained

$$a(bc + b_|c_|) + a_|b = 0.$$

This is in normal form; applying the normalization technique (B) $am + a_|n = 0$ which gives the result $mn = 0$, we have:

$$b(bc + b_|c_|) = 0,$$

which gives on expansion

$$bc = 0.$$

An alternative method is also given by Schröder which utilizes the transitive nature of $\nmid$. Since we have two subsumptions $a \nmid bc_| + b_|c$, and $b \nmid a$, applying the law of transitivity, we obtain the single subsumption: $b \nmid bc_| + b_|c$. Then applying Schröder's first technique (A), i.e., $a \nmid b$ implies $ab_| = 0$, we have as before $b(bc_| + b_|c)_| = 0$. Similarly, on taking the complement and multiplying out, we obtain $b(bc + b_|c_|) = 0$ and so $bc = 0$.

Another problem given by Schröder on page 529 of vol. 1 of the *Vorlesungen*, has as premises:

$$ab \nless cd_| + c_|d, bc \nless a_|d_| + a, di \text{ and } a_|b_| = c_|d_|.$$

Applying the principle that $a \nless b$ implies $ab_| = 0$; we get

(1) $$ab(cd_| + c_|d)_| + bc(ad + a_|d_|)_| + a_|b_|(c_|d_|)_| + c_|d_|(a_|b_|)_| = 0$$

Note that $a_|b_| = c_|d_|$ is interpreted as $a_|b_| \nless c_|d_|$ and $c_|d_| \nless a_|b_|$. From here Schröder gives just one other equation:

(2) $$a_|b_|c + (abc + a_|bc + a_|b_|)(abc_| + abc + ac_| + bc_|) = 0$$

before reaching the result:

(3) $$abc + a_|b_|c = 0$$

What are the intervening stages? On taking the complements in equation (1) we obtain:

(*) $$ab(cd + c_|d_|) + bc(ad_| + a_|d) + a_|b_|(c + d) + c_|d_|(a + b) = 0.$$

Multiplying out (*):

$$abcd + abc_|d_| + abcd_| + a_|bcd + a_|b_|c + a_|b_|d + ac_|d_| + bc_|d_| = 0.$$

Collecting up the terms containing d and $d_|$:

(2) $$(abc + a_|bc + a_|b_|)d + (abc_| + abc + a_|b_| + ac_| + bc_|)d_| + a_|b_|c = 0$$

The first two terms are in normal form for d, and applying the principle $am + a_|n = 0$ which gives the result $mn = 0$, we have:

$$(abc + a_|bc + a_|b_|)(abc_| + abc + a_|b_| + ac_| + bc_|) + a_|b_|c = 0.$$

Expanding and eliminating the complements $aa_| = 0$ etc. gives the required result:

(3) $$abc + a_|b_|c = 0$$

Peirce did not make any use of Schröder's problem-solving method. He disliked his emphasis on equations which, as can be seen clearly from the above two examples, was strongly featured. He wrote:[6]

---

[6]See [**84**, p. 450].

Somewhat intimately connected with the question of the relation between categoricals and hypotheticals is that of the quantification of
the predicate. This is the doctrine that identity, or equality, is the
fundamental relation involved in the copula. Holding as I do that the
fundamental relation of logic is the illative relation, and that only in
special cases does the premiss follow from the conclusion, I have in
a consistent and thoroughgoing manner opposed the doctrine of the
quantification of the predicate. Schröder seems to admit some of my
arguments; but still he has a very strong penchant for the equation.

Although subsumption relations are used initially to express the premises, these relations are quickly converted to an equational form to complete the problem solving.
Because of this, he considered Schröder an original and interesting proponent of the
algebra of classes following on from Boole rather than De Morgan. The influence of
Schröder on Peirce is negligible. It is first seen when Peirce noted favorably Schröder's
use of duality in [**153**] and compared this with the practice of geometricians. Many of
Schröder's formulae are given and he also stated that Schröder had not previously read
either De Morgan, Jevons or himself. By the time Schröder's main work was published
in the 1890s, Peirce had already developed his quantification theory. He disagreed with
Schröder's logic in many instances as will be seen in the next section, and by then he
had moved away from an algebraic approach to a graphical representation of logic in
the form of logic diagrams.

6.3. Schröder's Propositional Logic

Like Boole and Peirce prior to 1885, Schröder held that the truth of a proposition
was equivalent to its assertion and that there was a period of time, *Zeitpunkte*, during
which its truth could be asserted. Peirce however, although he returned to the former
argument and incorporated it into his existential graphs of 1897, repudiated the latter
argument.[7]. Schröder's formalist view, emphasizing the form of the axioms rather than
the interpretation of the symbols, which he shared with Peirce, led him to confuse
propositions and classes. As Grattan-Guinness writes:[8]

---

[7]See [**159**, p.216].
[8]See [**65**].

...he even continued to use the lower case letters of his calculus of classes to denote propositions. Indeed, in one part or another of his *Vorlesungen* he made the overworked symbol '0' denote the empty class, the empty relative, a contradiction, falsehood (in association with '='), and zero!

Schröder defined the empty class on page 188 of *Vorlesungen* vol. 1 as:

Definition (2_x) der *identischen Null* dadurch, dass wir die Subsumption $0 \{ a$ als eine *allgemeingültige*.

By defining the empty class in this way, Schröder showed that he was confused about his conception of the empty class, because with regard to his extensionalist view of classes, the empty class is literally nothing. Comparing this with the definition that we have seen earlier in Chapter 4, it is clear that Peirce also took an extensionalist view and thought of the empty class as 'nothing' or nonexistence. This is evident from the principle of the excluded middle which is expressed in *Logic of Relatives*, as $x+, n^x = 0$. However confusingly, Peirce also uses the same symbol to represent the zero relative which denotes WHAT EXISTS IF, AND ONLY IF, THERE IS NOT _____, 0 being the zero relative term such that $x + 0 = x$ and $x, 0 = 0$ vanishing when x exists, and not vanishing when x does not exist.

Moreover when used in propositional logic, Peirce equated 0 with 'falsehood' and 1 with 'truth'. This was of course a Boolean legacy[9] which made it impossible to distinguish between truth and tautological truth. So far Schröder agreed with Peirce in his use of 0. He also used this symbol for 'falsehood' and so confused theory with metatheory. As Grattan-Guinness writes:[10]

His views on truth and falsehood ...included a continuation of his confusions about the empty class, for, whereas he wrote $\overset{\bullet}{1}$ to represent truth in order to distinguish it from the symbol '1' for the universal class, he usually wrote '0' rather than $\overset{\bullet}{0}$ to represent falsehood because he thought that, like the empty class, it was 'an empty thing' (*ein leeres*). In this case the formalist confusion prevented

[9]See Chapter 3.
[10]See [**65**, p. 114].

him from realizing that '$A = 0$' (where 'A' represents a proposition) is doubly *interpretable*. For if '$= 0$' is taken to be a collective symbol, then it denotes the property of falsehood and '$A = 0$' asserts, in the metatheory, that A is false. But if '$= 0$' is taken to be a concatenation of '$=$' and '0', then '$=$' represents equivalence, '0' denotes a contradiction, and '$A = 0$' asserts, in the object theory, that A is equivalent to a contradiction.

Peirce however, was to develop the concept of 0 even further. In [**158**, p. 214], the question of truth values for propositional logic is dealt, with in a completely different way when he introduces the truth-values **v** and **f**, where $x=f$ means that the proposition x is false. He adopts the view here that every proposition is either true or false with no intermediate values. In this paper yet another method of asserting the truth of a proposition is given: the very writing of the symbol denotes the proposition is true. The fact that a proposition is false is asserted by writing a line over it. This now leaves the empty class 0 free from the ambiguous treatment mentioned earlier. In this way Schröder's failures concerning the empty class are resolved by Peirce. Schröder's use of different orders of universes or *mannigfaltigkeit* meant that many of the set-theoretical paradoxes were simply not expressible in his logic. This concept was derived from Peirce and Mitchell. The latter had introduced this in [**128**].

In the following passage[11] Schröder distinguishes between the two different manifolds 1 and 1̇,

> Zu ihrer bessern Unterscheidung von der bisherigen, eine räumliche Mannigfaltigkeit darstellenden oder auch im Klassenkalkul verwendeten (und auch noch fernerhin in dieser Wise zu verwendenden) 1, möge die Eins, als Symbol der Ewigkeit gedeutet mit einem Tupfen versehen ... 1̇

or

---

[11] See [**184**, p. 5].

To better distinguish from previous [definitions], a spacial manifold is depicted also applying to the calculus of classes as 1, a symbol which is interpreted as Eternity when shown with one dot ... $\overset{\bullet}{1}$

Coexisting with his extensionalist view of classes, Grattan-Guinness points out that his pure or 'reine' manifold was composed of classes of elements distinguished by some intensionalist property. These were elements of a derived or *abgeleitete* manifold, as opposed to the usual *gewöhnlichen* manifold composed of individuals. Solving for domain x the equation:

$$x + b = a$$

therefore $x = ab_| + uab =: a/b$ where u is an arbitrary domain. The elements of a/b belong to the derived domain. Grattan-Guinness also notes[12] that Schröder's subsumption is inadequate here as there should be two symbols for subsumption, one for the original and one for the derived manifold.

6.4. Schröder's Logic of Relatives

In Volume 3 of *Vorlesungen*, Schröder concentrated on the algebra of binary relatives, *Algebra der binaren Relative*. He fulsomely praises Peirce in his foreword and went on to cite Peirce's dual relatives on page 3. The fact that he does not consider relations of higher order in any great detail shows clearly the Peircean influence. Peirce himself always maintained that relatives of higher order could be constructed from dual and triple relatives. It is also no coincidence that the three relative operations included by Schröder, i.e., those of relative multiplication, relative addition and taking the converse, are exactly the three elementary operations used by Peirce.

Peirce's notation is also used in the introduction of elementary relatives of the form A: B when Schröder used a matrix or "block" formation similar to the matrices introduced by Peirce in *Logic of Relatives* who was himself influenced by his father

---

[12]See [**74**, Sec. 4.4].

Benjamin Peirce's work on linear associative algebras. [**186**, p. 10] has:

$$
\begin{array}{lllll}
A:A, & A:B, & A:C, & A:D, & \ldots \\
B:A, & B:B, & B:C, & B:D, & \ldots \\
C:A, & C:B, & C:C, & C:D, & \ldots \\
D:A, & D:B, & D:C, & D:D, & \ldots
\end{array}
$$

However his more usual formulation of a relation is in the form of an ordered pair or 'Elemente-paar' which he defines as

$$
\begin{aligned}
1 = \Sigma_{ij}\, i:j = {} & A\!:\!A + A\!:\!B + A\!:\!C + \ldots \\
& + B\!:\!A + B\!:\!B + B\!:\!C + \ldots \\
& + C\!:\!A + C\!:\!B + C\!:\!C + \ldots
\end{aligned}
$$

where ij represents an individual binary relative. In using ordered pairs rather than the domain class of the relative, Schröder was following the later Peirce position as introduced in [**157**]. In fact it is clear that this paper and also [**158**] which introduced the quantifiers, had a great effect on Schröder's work particularly on his logic of relations and his propositional logic. This can be seen throughout the *Vorlesungen* volumes in the references to Ladd-Franklin and Mitchell. Dipert writes:[13]

> Peirce, especially took the almost wholesale adoption of his approach by Schröder in the *Vorlesungen* (quantifiers, the subsumption operator, and most importantly, the theory of relatives) as confirmation of the merits of his own theories.

The ordered pair formulation emphasizes the relation and is more convenient for quantification than the "relative" definition of *Logic of Relatives*, which emphasized the domain class of the relation.

Unlike Peirce, Schröder used different universes to distinguish between different types of relations, e.g., 1^1 denotes the universe of unary relations, 1^2 denotes the universe of binary relations etc.[14], although he now deployed the word *Denkbereiche* rather than *Mannigfaltigkeit*.

---

[13]See [**46**, p. 132].
[14]See [**186**, p. 12].

In his formulation of a binary relative, Schröder used the Boolean coefficients of Peirce although without his inequality symbol. He defined it thus:

(5) $$a = \Sigma_{ij}a_{ij}(i : j)$$

"where 'the 'Coefficient' a_{ij} (spoken a subscript ij), with which the element pair ij is ... associated or apparently multiplied, is limited on both sides by the Values 1 and 0."[15]

This manner of definition Peirce soon abandoned in favor of a superior notation identifying the coefficient with the relation[16], where the assertion of the relation ensures its existence as previously outlined in Chapter 5 above. By such means the relative coefficients with values 0 and 1 were rendered unnecessary. However the appearance of such relative coefficients in Schröder's work emphasizes the fact that his main source was [**120**].

6.5. The Peirce-Schröder Correspondence

In fact, Peirce and Schröder began to correspond in 1879, with Schröder receiving from Peirce a copy of *Logic of Relatives* in this year. He felt obliged to give Peirce credit for prior work done which had then appeared subsequently in his *Operationskreis*, 1887. Initially, relations between the two logicians were cordial. In the earliest letter extant dated 1 February 1890, Schröder tells Peirce that up to 1884, 'I rejoiced in receiving your communications." Peirce also adopted the *Operationskreis* as a text for his Johns Hopkins logic course.[17]

For the five years following 1885, there was no further correspondence and Schröder suspected that he had somehow offended Peirce. One bone of contention between Peirce and Schröder was Schröder's claim which he published in Vorlesungen that the distributive principle

$$(a + b) \times c \prec (a \times c) + (b \times c)$$

could not be proved from the definitions given in [**153**]. Peirce's proof for the distributive principle was later printed in [**99**, p. 300], but at the time he seemed to accept this correction at face value.

---

[15]Ibid.
[16]See [**158**].
[17]See [**97**, p. 206].

It is most unfortunate that much of the *Nachlass* of Schröder was lost including correspondence with Peirce, Russell and Peano. These papers were given to the *Archives of the Technische Hochschule* in Karlesruhe on Schröder's death. However they were later removed and kept in the basement of a building of the University of Munster, the site of the first, massive daylight bombing mission by the Allies in 1943. In a raid on March 25th 1945, the building where the *Nachlässe* of Frege and Schröder were stored, was completely destroyed. However a portion of Frege's papers had been transcribed by the time the archive was destroyed. This was not true of Schröder's papers.

The period from 1885, following Peirce's forced resignation from the Johns Hopkins University, his resignation from the U. S. Coast Survey and his purchase of a country estate at Milford, Pennsylvania (bought with the help of an inheritance from an aunt), was an unsettled and troubled time for Peirce. He wrote in a draft letter to Christine Ladd-Franklin on 29 August 1891:

> ... if Schröder's manner seems a little harsh toward me, that is more than excused by the manner in which I have neglected to write to him. He does not know, and nobody can begin to imagine, the difficulties under which I have labored.

This last point delicately hints at the fact that an extensive correspondence was something Peirce could not afford to support. He had worked for most of his life outside mainstream American mathematics. At first this was not a serious disadvantage, e.g., his time at the Geodetic Survey under the benevolent eye of his father. However due to serious personal inadequacies, such as his lack of self-control and social skills, he was never employed as a mathematics lecturer at Harvard unlike his older brother James; even his part-time lectureship at the Johns Hopkins ended because of his irregular (for the time) divorce and remarriage to Juliette Portelai. His inheritance should have made him independent but this was soon squandered on speculative ventures, bad investments and extensive rebuilding of his new mansion which he christened 'Arisbe'. He was to die here in a penurous state in conditions of cold and hunger [17].

In his final letter to Schröder, Peirce asked for his support in his application for a grant from the Carnegie Institution that was to result in the publication of his definitive work on logic. Peirce added:

I shall for my part, be warmly in favor of some of the money being employed to aid the completion of your great work.[18]

Unfortunately Schröder had already died so the letter was returned to Milford.[19]

Schröder's high opinion of Peirce was revealed when he wrote to Paul Carus, editor of *The Monist*, that Peirce's fame 'would shine like that of Leibniz or Aristoteles into all the thousands of years to come' and he invited Peirce to contribute to the third volume of *Vorlesungen*. This regard was evidently reciprocated. In 1896, Peirce believing he was on the point of death gave instructions that Schröder should have his logic manuscripts. However relations between the two logicians were later to deteriorate. Schröder introduced the term *symmetrical* for De Morgan and Peirce's *convertible*, and by 1897 Peirce had taken objection to Schröder's 'wanton disregard of the admirable traditional terminology of logic which "would result in utter uncertainty as to what any writer on logic might mean to say, and would thus be utterly fatal to all our efforts to render logic exact."[20] This seems a little hypocritical coming from one who often introduced novel terminologies himself.[21]

Of course, Peirce was in reality objecting to the fact that his own symbols and terminology were being replaced by Schröder's own. As he said "priority must be respected, or all will fall into chaos"[22], and later regretted the fact that he too had been guilty of the same crime when he renamed his master De Morgan's algebraic theory of relations, the logic of relatives.

6.6. Differences between the Logics of Peirce and Schröder

6.6.1. A Logic of Classes and/or Propositions. Schröder considered that one of the most important differences between his treatment and that of Peirce's was the clear distinction between the logic of classes and the logic of propositions. In his work, logical variables cannot apply to both classes and propositions. He used lower

---

[18]This references it to *Vorlesungen*, of which only three volumes were published in Schröder's lifetime.
[19]See [**97**, p. 207].
[20]See [**84**, p. 300].
[21]Martin Gardner writes of Peirce's "opaque style, his use of scores of strange terms invented by himself and altered from time to time" [**62**, p. 55–56].
[22]See [**158**, p. 286].   ı

case letters for classes and upper case letters for propositions. In contrast the logical variables of Peirce apply to both classes and propositions. Schröder was to write:[23]

>   ... every theorem/proposition holding good in the class-calculus also holds in the statement-calculus. But not vice versa.

By the statement-calculus Schröder means propositional logic. Peirce replied "What you call the statement-calculus is nothing but what the calculus of logic becomes when but a single individual forms the universe of discourse."[24]

Another instance of this important difference was when Peirce together with Ladd-Franklin argued that it was not justifiable to regard $xy + z$ as requiring fundamentally different treatment according as to whether x, y and z stand for terms or for propositions. Schröder had maintained that, when the letters represent propositions, it is not possible, as it is when dealing with classes, for x to be divided up between y and z.[25] Shearman also considered MacColl took the view that the symbols represented exclusively propositions. However it is the case that MacColl gave epistemological priority to propositions but symbols are not so restricted.

Another point of divergence between the logics of Peirce and Schröder is in their differing philosophical treatments of the transition from a logic of classes to a logic of propositions. For Boole the nature of the universe for secondary propositions (hypothetical or mixed such as IF A IS B THEN C IS D rather than primary propositions which were categorical such as ALL A IS B), changed from possible cases or circumstances as in *Mathematical Analysis of Logic* to durations of time for which the propositions were true as in *Laws of Thought*. Schröder in the second volume of the Vorlesungen, which was devoted to propositional logic, inclined towards the latter view. According to Dipert[26], Peirce in his later work harshly rejected this position, preferring instead to speak of possible situations; however this was not evident in his earlier papers, and as we have seen both Mitchell and Ladd-Franklin in [120], used the temporal concept of the universe of propositional logic, i.e., ∞ was used to signify a universe of possibility where propositions were valid for all of the time with subsets u of this universe where propositions were true for only some of the time.

---

[23]See [**97**, p. 224].
[24]Ibid. p. 229
[25]See [**190**, p. 16].
[26]See [**47**, p. 522].

In this, Peirce and his students from the Johns Hopkins University followed Boole in applying a temporal notation so that propositions are logically consistent if they are both true for the same moment of time. Peirce however moved away from this position and when Schröder also took the temporal view, Peirce criticized him severely. Dipert states:[27]

> In Boole's *Mathematical Analysis of Logic* the referent of a term in the secondary or propositional context is described as 'conceivable cases or conjunctures of circumstances,' while in his more famous 1854 *Laws of Thought,* they are identified with durations of time. Schröder, in the second volume of the *Vorlesungen,* 1890, which was devoted to propositional logic, followed Boole's second formulation in making the referents of propositional terms temporal entities, while Peirce harshly rejected this position, preferring instead to speak of "possible situations" as the referents of propositional terms.

Schröder for example speaks of a 'duration of validity' (Gltigkeitsdauer) which consists of individual time periods (Zeitpunkten). An important difference that occurs in Schröder's logic is its equational structure and according to Peirce, overemphasis on mathematical terms and constructions. He claimed:[28]

> It is Schröder's predilection for equations which motives his preference for the algebra of dual relatives, namely, the fact that in that algebra, even a simple undetermined inequality can be expressed as an equation. ... He looks at the problems of logic through the spectacles of equations, and he formulates them, from that point of view as he thinks, with great generality; but, as I think, in a narrow spirit. The great thing, with him, is to solve a proposition, and get a value of x, that is, an equation of which x forms one member without occurring the other. How far such an equation is *iconic,* that is, has a meaning, or exhibits the constitution of x, he hardly seems to care.

This paper also calls to attention "a mere algebraic difference" between the two logicians. Peirce restricted the possible values of his propositions to 0 and 1 so that every proposition is, during the limits of the discussion, either always true or always

---

[27]See [**47**, pp. 512–522].
[28]See [**159**, pp. 284–285].

false. For Schröder a class variable has the possibility to take as a value a class which is neither null nor identical to the class of everything. Schröder calls these additional possible denotations of a class symbol *Zwischenwerte*: middle, or between values. Peirce blamed himself for this divergence. He attributed it to his notation $\Sigma_i\Sigma_j l_{ij} > 0$ which means that SOMETHING IS A LOVER OF SOMETHING, first developed in [**153**]. He accused Schröder of taking this literally and moving away from the logic of Boole. However although restricting himself to two values for his logic of classes and propositions, he was able to express *umbrae* or inbetween values in his relational logic as early as *Logic of Relatives* when he introduced the relative term 0^x which vanishes when x is non-empty and does not vanish otherwise. $0^{RA} = 0$ then expresses the fact that if RA is the class of rotten apples then some but not necessarily all apples are rotten. Dipert comments:[29]

> This is workable (or so Peirce thinks) because the semantics for relations and functions is less restricted than the two-valued semantics for the propositional and monadic predicate (= categorical) calculus. To explore this method would take us far afield into Peirce's theory of relations, and we must here content ourselves with the observation that the nonrelative calculus is for Peirce, but not for Schröder and most Booleans, inadequate for properly expressing "umbral" statements.

However, this seems to be at odds with Peirce's rejection of the Reduction Hypothesis and his claim that the categorical and hypothetical calculus are the same. Dipert suggests that this restriction of class and propositional values to 0 and 1 is merely an algebraic convenience:[30]

> This might mean that although Peirce's interpretation is a convention, it is a *preferable* convention. For example, one might argue that the assumption of two values is the minimum necessary assumption in order to have a formal system with any content at all. Thus we must assume, thinking categorically, that something exists; yet the assumption that more than one thing exists, and the incorporation of this possibility into the basic calculus, is to laden the calculus with a content discoverable only through experience.

---

[29]See [**43**, p. 589].
[30]*Ibid.*

These differences between the two logicians sprang from their differing philosophical positions. To Peirce, iconicity was paramount in logic and he was clearly aware of the distinction between mathematics and logic. Schröder as a mathematics lecturer believed that algebraic logic was a model which would express formal algebra, and so justified Peirce's accusation of being too mathematical. Indeed his emphasis on duality exceeded anything that Peirce had attempted. For example Schröder proved that the complement of $(ab)_|$, is $(a_| + b_|)$ and that of $(a + b)_|$ is $a_|b_|$ (where $a_|$ is the complement of a) - in the *Vorlesungen* the proposition appears as No. 36 on page 352 of volume 1.

This follows from a previous theorem of Schröder's namely Theorem §30 which states $ab + (a_| + b_|) = 1$; from this we get $(ab)_| = (a_| + b_|)$. Peirce did not present his theorems in this way. This layout came from the projective geometers of the nineteenth century who used double columns for dual equations relevant to points, lines and planes.

Notational differences aside, Peirce criticized Schröder severely in his second Harvard Lecture of 1903. He claimed that Schröder:

> was too mathematical, not enough of the logician in him. The most striking thing in his first volume is a fallacy. His mode of presentation rests on a mistake and his second volume which defends it is largely retracted in his third [and] is one big blunder.

This fallacy concerned the different positions the respective logicians took over the hypothetical/categorical debate. Typically, Peirce was later to praise the *Vorlesungen*, particularly volume 3 on the logic of relatives which followed his own work in later published papers such as:[31]

> ... in 1895, Schröder devoted the crowning chapter of his great work (*Exakte Logik*, iii. 553-649) to [the logic of relatives and] its development ...

and in [**159**, p. 267]:

---

[31]See [**84**, p. 389].

...the man who sets up to be a logician development without having gone carefully through Schröder's *Logic* will be tormented by the burning brand of *false pretender* in his conscience, until he has performed that task.

6.6.2. The Reduction Hypothesis. The hypothetical/categorical debate centered on the question: What is the relation between the logic of classes and the logic of propositions? The Reduction Hypothesis claimed that hypothetical propositions could be reduced to categoricals, i.e., the hypothetical IF A IS B THEN C IS D, e.g., IF IT THUNDERS, IT LIGHTENS' can be reduced to the categorical ALL A IS B, e.g., 'ALL THE OCCASIONS IN WHICH IT THUNDERS ARE OCCASIONS IN WHICH IT LIGHTENS.

Peirce however, supported a different form of the Reduction Hypothesis, i.e., that categoricals are in fact hypotheticals:[32]

The categorical proposition, EVERY MAN IS MORTAL is but a modification of the hypothetical proposition, IF HUMANITY, THEN MORTALITY.

In his later papers, he came to a third version in which categoricals and hypotheticals could be reduced to a third kind of statement.[33]. Peirce set out his position thus:[34]

The forms $A \prec B$, or A implies B, and $A \not\prec B$, or A does not imply B, embrace both hypothetical and categorical propositions'

Peirce interpreted the conditional *de messe* IF Ai IS TRUE THEN Bi IS TRUE, where i denotes the actual state of things, as EITHER Ai IS NOT TRUE, OR Bi IS TRUE.

In contrast, Schröder held that hypotheticals could be reduced to categoricals involving time, e.g., ALL THE TIMES WHEN IT THUNDERS ARE TIMES WHEN IT LIGHTENS' but rejected the Reduction Hypothesis because he thought the calculus appropriate to hypotheticals contained some rules inapplicable to categoricals generally. He tried to explain his point of view and overcome the "mental idiosyncrasy" of Peirce's opposing position when he wrote on 2 March 1897:

---

[32]See [**83**, p. 710].
[33]See [**43**, p. 575].
[34]See [**84**, p. 175].

The very simple fact being: That every theorem/proposition holding good in the class-calculus also holds in the statement calculus. But not vice versa. There are many formulae holding in the latter, but not in the former.

Peirce replied on 7 April 1897:

What you call the statement-calculus is nothing but what the calculus of logic becomes when but a single individual forms the universe of discourse ... I said to myself "When he comes to the logic of relatives and has to deal with indices, he will see that there is no difference between the calculus of statements and that of classes." How it is you fail still to see the identity of the two, long puzzled me.

By this he meant that he held the formal structure of the algebra to be of utmost importance. It is the interpretation of the symbols that provides the meaning and to Peirce the symbols can either signify classes to provide predicate logic or propositions to give a logic of propositions. This shows the algebraic influence of both Boole and his father Benjamin Peirce.

The crux of the matter is, as argued by Dipert, that Peirce understood that the logical relations of conditional, inclusion and logical consequence are deeply similar. He writes:[35]

Peirce is interested in developing a formal calculus whose intended interpretation is indifferently propositional logic, the logic of classes, or a metalogical theory of logical consequence, depending on what the reference of the terms is taken to be.

Schröder, taking a different view, not only wished to differentiate between the logic of classes and the logic of propositions (although as we have seen he was not always successful) but was moreover concerned with categorizing individuals, classes, dual relatives, triple relatives etc. This can be seen in the next section.

---

[35]See [**43**, p. 581].

6.6.3. The Universe of Relative Terms. In order to distinguish between his logical terms, Schröder introduced a number of different universes – a universe for individuals, a universe for propositions, a universe for pairs etc. Peirce however, following Mitchell, used only a universe for classes 1 and a universe for relative terms 1. Schröder took up the matter strongly. He wrote on 7 March 1892:

> ...it is absolutely indispensable - for the sake of *reconciling* your definition of the individual (say dual or binary) relative as *signifying* (or meaning) the *relate*, with the main theory of dual relatives: to *distinguish from the outset different realms or universes* ("Denkbereiche"), which I denote by $1^1, 1^2, 1^3, \ldots$ (rejecting your ∞ [written as 1^2], and replacing your relative moduli $\mathbf{n}$ and 1 by 0^1 and 1^1).

In this he was influenced by the different dimensions of Mitchell, as he was always careful, unlike Peirce, to distinguish constants 1 for the universe of individuals, $1'$ or 1^1 for the universe of all propositions, $1^{(2)}$ for the universe of all pairs, and so on[36]. He did not use the word 'universe' but rather *Mannigfaltigkeit* or manifold. Church asserts that Schröder anticipated Russell's simple theory of types.[37] However Grattan-Guinness has shown[38] that Schröder's arguments for a type theory rested on "dubious assumptions connected with his unclear characterisations of individuals and classes." The point being that the similarity is superficial in that Betrand Russell used sets and Schröder the part/whole theory of classes, so that no real comparison is possible.

6.7. Similarities

Christine Ladd-Franklin noticed the similarity between the presentation of Schröder's volume 1 of the *Vorlesungen* and Peirce's own method of working, i.e., to establish all the formulae by analytical proofs based upon the definitions of sum, product, and negation, and upon the axioms of identity and the syllogism. These proofs were the same as those given by Peirce, but with alternative proofs inserted and an occasional difference in the method used. Schröder also shared with Peirce his extensive output. Peirce's life aim was to produce a series of volumes on his logic - an

---

[36]See [**188**, p. 8].
[37]See [**27**, p. 151].
[38]See [**65**, p. 125].

aim never realized. However Christine Ladd-Franklin compared Schröder unfavorably with Peirce, accusing Schröder of being:[39]

> unnecessarily diffuse ... and discursive to the last degree

whilst at the same time praising Peirce's work as a model of abstractness and brevity.

Schröder together with Jevons and Peirce, used the non-exclusive form of addition in direct contrast to Boole who always used + on the understanding that the classes to be joined are mutually exclusive. One advantage of using non-exclusive addition is the ease of producing formulae involving negatives. In fact, Peirce and Schröder used operations in a similar way. Peirce's main logical operation was that of illation $\prec$ signifying inclusion. He favored this as a wider and therefore a more basic and foundational operation than the equality of Boole. Schröder also placed his version of this operation called 'subsumption' $\{$ at the foundation of the calculus of classes not the equality relation as mentioned earlier.

As well as the basic copula of logic, in another important development of logic,i.e., the theory of quantification, Schröder also followed Peirce in his use of Σ and Π as abstract operators which are variable-binding. Dipert writes:[40]

> By 1885, both authors regard the use of some variable-binding operators as necessary for the proper formulation of the relational calculus.

Although Schröder only used the quantifiers separately in multiple additions or multiplications, by his third volume there was explicit use of mixed types, e.g.,

$$\Sigma_u \Pi_v A_{u,v} \{ \Pi_v \Sigma_u A_{u,v}.$$

Schröder's novel introduction of hierarchies of manifolds in *Vorlesungen* I, p. 243 ff mirrors Peirce's idea of dimensions which he defined in (Baldwin 1911) as 'an element or respect of extension of a logical universe of such a nature that the same term which is individual in one such element of extension is not so in another. ' As we have seen in Peirce's algebraic logic an individual can also be expressed as a monadic relative term. Similarly, in Schröder, classes or domains in one manifold are considered as

---

[39]See [**111**, p. 126].
[40]See [**45**, p. 55].

individuals in another manifold. He briefly considered two kinds of sign for subsumption, one for the original and one for the derived manifold. However, Schröder did not pursue this thought which showed that he did not have a clear idea of a distinction between elementhood and class inclusion. He had no need of this distinction because like Peirce he was working in the part/whole theory of classes.

Another common thread found in the work of both Peirce and Schröder was their lack of interest in the axiomatisation of their quantified relational predicate calculus in a systematic way. Peirce himself complained that Schröder's third volume of *Vorlesungen* had no obvious significance or direction. Dipert's view is that Peirce and Schröder lacked the unification of a single goal, i.e., a logicist program to derive the Peano-Dedekind postulates that drove Frege, Russell and Whitehead. He states:[41]

> The unarguable fact that the Peirce-Schröder calculus was not actually applied by its discoverers to the foundations of mathematics seems to be more a consequence of the fact that Peirce and Schröder were not logicists, rather than of substantive flaws in their calculus

It should however be noted that Schröder was a logicist in the sense that he viewed algebraic logic as a model for formal algebra and regarded arithmetic as part of a general logic.[42]

Consideration of Peirce's influence on Schröder is of vital importance when looking at the historical development of algebraic logic which was eclipsed by the mathematical logic that followed it. However our modern notation for predicate logic came from Peirce's work through Schröder and mostly from Peano and not from Frege "whose work was carefully read only much later and whose notational system was never used by anyone else."[43] The Peirce-Schröder calculus was portrayed as purely algebraic, without the variable-binding operators Peirce regarded as essential and which Schröder also used. The development of the theory of relations in *Principia Mathematica* was reinvented by Bertrand Russell, though Wiener showed it was largely unnecessary in the second chapter of his dissertation.[44]

---

[41]See [**45**, p. 52].
[42]See [**136**, p. 357].
[43]See [**47**, p. 529].
[44]See [**65**].

Peirce and Schröder felt united by a common aim – to promote algebraic logic rather than the mathematical logic of those who used the notation of Peano and Cantor. Schröder wrote to Peirce on 2 March 1897:[45]

> By the bye you had better not [emphasize] the comparatively trifling divergences of our systems of notation in view of the contrast with the latter of the one, unanimously employed by those most active Italian investigators [probably Peano], which is, at least with regard to relative notions, so very inferior to ours. I have so to say to stand out nearly alone against them all whereby "the Good" again and again proves to be an enemy of "the Better" – as is averred by the proverb: Das Gute ist des Bessern Feind.

Peirce himself felt that the differences in their opinions was healthy – the sign of a living science and not a dead doctrine. He wrote:[46]

> Professor Schröder and I have a common method which we shall ultimately succeed in applying to our differences, and we shall settle them to our common satisfaction; and when that method is pouring in upon us new and incontrovertible positively valuable results, it will be as nothing to either of us to confess that where he had not yet been able to apply that method he has fallen into error.

---

[45]See [**97**, p. 224].
[46]See [**159**, p. 287].

CHAPTER 7

Summary and Conclusions

7.1. Introduction

In this chapter the main influences and achievements of Charles Sanders Peirce are discussed, and a brief survey of the development of Peirce's logic is provided, starting from his own improvement of Boolean algebraic logic, moving through to the powerful algebraic logic of *Logic of Relatives* to his discovery of quantificational logic and the final development of a graphical form of logic in the existential graphs. The work of Peirce scholars will also be noted, and in particular the influence of Peirce's algebraic and quantificational logic on later work in logic.

7.2. The Three Main Influences on Peirce's Logic

7.2.1. Early Influences: Benjamin Peirce. The most important early influences on Peirce's algebraic logic came from three sources: Benjamin Peirce, George Boole and Augustus De Morgan. Let us first consider Benjamin Peirce (1809–1880), who was the most important American algebraist of his time and who deeply influenced his son's work. Jacqueline Brunning has shown that it is Benjamin's general notion of multiplication developed for his algebras is that used by Charles for the operation of relative multiplication in *Logic of Relatives*.[1] She has also conjectured that since Charles adopted the multiplication schema of his father, it seems safe to assume that he was inclined to show that relatives exhibit properties similar to the units of a linear algebra. He was certainly to express absolute terms as a linear combination of individuals and relative terms as a linear combination of elementary relatives.[2]

---

[1] See [**19**].
[2] See [**149**, pp. 370–371].

James Van Evra considers Peirce's important paper of 1870 on relatives, and shows that the connection with his father Benjamin Peirce's *Linear Associative Algebra* as published in 1807 is evident in examples of some of the algebras appearing in *Logic of Relatives*.[3] Peirce claimed that the algebras in *Linear Associative Algebra* have a representation in his logic and later went on to show that all algebras derive from those in *Linear Associative Algebra*. Van Evra points to the fact that the algebras of *Linear Associative Algebra* are abstract and do not always obey the commutative law (the algebra representing Hamilton's quaternions being one example), which would surely have influenced Peirce in loosing the ties to the traditional algebraic analogies. It is also clear from looking at *Linear Associative Algebra* that Benjamin Peirce emphasized the laws of algebra rather than the meaning of the symbols, i.e., form rather than matter. Peirce was to endorse his father's philosophy that any necessary conclusions were to be made from form alone.[4] This emphasis on form with the interpretation being but a secondary consideration was facilitated by Benjamin's strong religious beliefs.[5]

The close connection between *Logic of Relatives* and *Linear Associative Algebra* both written in 1870 is further emphasized by the fact that the more complex algebraic portion of *Logic of Relatives* culminates in the appearance of some of the algebras from *Linear Associative Algebra* and one of the few applications cited by Peirce in using his algebraic logic is precisely to express the algebras from *Linear Associative Algebra*. Peirce was influenced by the purely formal approach of *Linear Associative Algebra*, in which algebras did not obey the commutative law, to free his logic from the restricted operations that Boole followed in his mathematical analogies. In *Linear Associative Algebra*, Benjamin Peirce gave a broad definition of mathematics as the science which draws necessary conclusions and extends algebra to cover formal mathematics. In the same way, Peirce extended the algebraic analogy to cover all of formal logic. This provided the rationale for some of his more obscure mathematical analogies as found in *Logic of Relatives*, which came from the inspiration of his father Benjamin rather than the more restricted Boolean analogies. However a clear distinction between the

---

[3]See [**199**, pp. 147–157].

[4]See [**72**, p. 35].

[5]The argument runs something like this. Symbolical algebras are reflections of the Divine Mind and so must have some physical reality. Both Nature and Mathematics originated from God, so worrying about the applicability of such algebras is pointless.

two separate disciplines of logic and mathematics was maintained by Benjamin: mathematics draws conclusions, logic theorizes about it. In this he was influenced by the arguments of his son Charles rather than the other way round.[6]

Linear Associative Algebra is essentially a classification of associative algebras by their units. These algebras are expressed by means of a matrix grid of the algebraic expressions that result from a process of multiplication in which units are either idempotent, nilpotent or expressible as some linear combination of other units. *Linear Associative Algebra* consists of a taxonomy of all possible linear associative algebras in the form of their multiplication tables for systems of up to six units resulting in a definition of one-hundred and sixty-three algebras and six subcases. Previous commentators have concentrated their efforts in proving and extending Benjamin Peirce's results by using other mathematical techniques (this was shown in section 2.10). We have seen that Hawkes sought to revive and publicize *Linear Associative Algebra* in 1902 by attempting to solve and relate the work to the number systems of G. Scheffers. Taber in 1904 attempted a similar task but used scalar function theory and criticized Hawkes for using group theory rather than an algebraic method. By 1907, J. B. Shaw was able to review the latest developments in linear associative algebras both in terms of group theory and matrix theory.

Grattan-Guinness notes[7] the effect that Peirce's algebras had on the development of algebraic logic at Harvard, particularly on the philosopher Josiah Royce (1855–1916) who reformulated a very general theory of collections due to the English mathematician A. B. Kempe in an algebraically symmetric way, and continued to analyze symmetric and asymmetric relations. Later, Peirce's use of multiplication tables was applied to mathematical logic by Royce's colleague H. M. Sheffer (1882–1964). Another interesting link between linear associative algebras and model theory arose when L. E. Dickson (1874–1954) sought to define such an algebra by independent postulates[8]

Chapter Two traces the algebraic methods used to produce many of the multiplication tables for the algebras in *Linear Associative Algebra* (1870) and supplies any algebraic reasoning omitted by Benjamin. Errors that have never before been corrected

---

[6]See Section 7.4.1.
[7]See [**72**, p. 600].
[8]See [**42**].

have now been indicated. No such analysis of the algebraic methods has previously
been attempted.

7.2.2. Early influences: George Boole. Much of Charles Peirce's early innovative developments in algebraic logic came primarily from study of the logic of George
Boole. In 1847, Boole sought to express logic, which up to that time had been the
province of philosophers and couched in terms of Aristotelian syllogisms, as algebraic
equations. However as Van Evra points out there are limits to which this mathematical
analogy can be extended and Peirce and Boole reacted to these limitations in different
ways.[9] On the one hand, Boole confined the analogy to elementary algebraic operations and their inverses with values taken by the logical variables restricted to 0 and 1.
This still presented problems with certain expressions that had no logical interpretation. He also used his 'general method' that essentially dealt with any uninterpretable
logical expressions by accepting as valid any process which started with logically interpretable premises and finished with a logically valid inference even though the steps
on the way may not have had a logical interpretation – a case of the ends justifying the
means, as it were.

But to Peirce, these uninterpretable expressions were a failing of Boole's system
and in order to correct this he provided symbols and interpretations for all algebraic
expressions resulting from the use of Boole's development theorem. Peirce considered
that it was the aim of mathematics to pass from premise to conclusion in a swift and
direct manner but judged a system of logic by the clarity of its individual steps. Therefore he could not agree with Boole over uninterpreted expressions. Emily Michael
points out that his view of logic was essentially semantic in that logic should serve as
an analysis of natural language.[10] In this respect, he disagreed with Boole, as there
would be no place for uninterpreted expressions in his logic. In his interpretations,
Boole had alternative meanings for his logical terms according to whether they were
contained in categorical or particular propositions (SOME X IS Y), or in hypothetical
propositions (IF A IS B, then C IS D). In *Laws of Thought*, when considering hypothetical propositions, Boole interpreted his symbols as the periods of time for which such
propositions were true. This was to reappear in later papers by Peirce. It was the young
Charles Peirce's dissatisfaction with Boole's treatment of categorical propositions that
led to him combining De Morgan's theory of relations with Boole's calculus to form
his own algebraic logic in the 1870 paper *Logic of Relatives*. Michael also notes that

---

[9]See [**199**, p. 151].
[10]See [**127**].

Peirce was dissatisfied with Boole's treatment of universal as well as conditional or hypothetical propositions, and saw the need to distinguish between Boole's mathematical symbols and logical symbols.

7.2.3. Early influences: Augustus De Morgan. It was a series of four papers on the syllogism written by De Morgan between 1846 and 1860 setting out his theory of syllogistic reasoning that captured Peirce's imagination and led to his development and extension of De Morgan's work on relations. In the analysis of this series of De Morgan's papers in Chapter 3, it was shown that one of the main points of influence on Peirce by De Morgan was the introduction of a universe of discourse. First introduced in 1846, this restricted the universe to cover the scope of the proposition or term under discussion. De Morgan used a limited universe because he wished to use complementation and the complements of well behaved and defined classes might not be well defined. This concept was to lead to the 'universe of discourse' introduced in De Morgan's "Formal Logic," of 1847. Dipert has shown[11] that Peirce was able to avoid the paradoxes that plagued the later set theorists by using De Morgan's 'universe of discourse' and avoid such concepts as 'the class of all classes'. Peirce in his later work was to construct a logical theory (his existential graphs being one such system), which had a number of universes or dimensions and such that a term that is an individual in one universe may not be an individual in another.

One other area of interest to De Morgan, which was to later influence Peirce, was the concept of quantification. Having embroiled himself in a dispute with Hamilton over who had the precedence in introducing the quantification of the predicate, e. g., SOME X IS SOME Y, De Morgan came to realize that his numerically definite syllogisms, in which he assigned a numerical value to subject or predicate, had little connection with Hamilton's more general 'some' or 'all' He was also concerned that categorical propositions such as EVERY HEAD OF A MAN IS THE HEAD OF AN ANIMAL could not be deduced syllogistically from EVERY MAN IS AN ANIMAL. It is interesting to note the use of relations in this example; indeed the concept of quantification in general and this particular example proved to be an inspiration for Peirce when he came to consider the next development of his algebraic logic - that of a logic with quantifiers.

In his description of relations De Morgan gave equal consideration to subject and predicate such as $X..LY$ where X (the subject) is an L of Y (the predicate). This shows his focus on the relation L as opposed to the domain class X or range Y of the relation.

---

[11]See [**49**]

He considered that composition of relations was a form of multiplication of classes but surprisingly omitted any consideration of the addition of relations. Michael's paper [**126**] on Peirce's early study of the logic of relations from 1865 to 1867 supports the view that Peirce discovered the logic of relations independently of De Morgan. Even if this is the case, then De Morgan's algebraic symbols certainly influenced its later development.

Consider for example De Morgan's subscript and superscript notation for universal and particular quantification. He used a subscript prime and parenthesis notation ,(to represent EVERY and a superscript prime and parenthesis (' for the existential quantifier ONE OR MORE. This subscript and superscript notation was to reoccur in Peirce's revised representations of quantification in *Logic of Relatives*. In fact Peirce was to delay publication of *Logic of Relatives* to include a superscript notation (used in the operation of involution) to express quantification, having just read De Morgan's "On the Syllogism IV".

Although initially influenced by the logic of George Boole, Peirce quickly became dissatisfied with the Boolean treatment of categorical propositions SOME X IS Y. Augustus De Morgan before him had realized the inadequacies of syllogistic logic and claimed that some way of representing relations other than the identity relation was needed. His theory of relations involved expressing inferences in logic in terms of composition of relations

$$X..LY \quad X \text{ is an } L \text{ of } Y$$
$$X.LY \quad X \text{ is not an } L \text{ of } Y$$

De Morgan issued a challenge: Deduce EVERY HEAD OF A MAN IS THE HEAD OF AN ANIMAL from EVERY MAN IS AN ANIMAL. He claimed that this was impossible unless relations other than the identity relation were used. Charles Peirce found this challenge irresistible. He had already concluded that ultimately the Boolean system was inadequate to treat mathematical propositions without introducing a theory of relations and went on to combine De Morgan's theory of relations with Boole's calculus to form his own algebraic logic in the 1870 paper "Description of a Notation for the Logic of Relatives" also known as *Logic of Relatives*. In this work, Peirce extended even further the algebraic operations in the search for fruitful logical results. These included differentiation, the Binomial Theorem and logarithms. In particular the process of 'logical differentiation' is introduced by Pierce as a direct analogue of mathematical differentiation. This section has long puzzled Peirce scholars mainly because although the algebraic techniques used are provided, any English interpretations for the logical terms are missing, so throwing a mysterious veil over the whole process. Not only are such interpretations suggested by other unclear concepts in *Logic of Relatives,* such as the 'number' of a class and the meaning of coefficients in his logical algebra are clarified.

Peirce claimed to have discovered the theory of relations independently of De Morgan, but even his initial work on relations was inspired by De Morgan's relational examples. He also extensively revised *Logic of Relatives,* in light of the publication of De Morgan's "On the Syllogism, IV." However, note that Peirce's algebraic logic is the 'logic of relatives' not De Morgan's 'logic of relations'. The emphasis changed in 1882 when he defined a relative term as a class of ordered pairs, i.e., as what we recognize today as a relation. However intensional interpretations were still admitted by Peirce.[12] This change is most clearly stated in his 1897 paper "The Logic of Relatives" as the change from classes to relations or as Peirce wrote:[13]

> The best treatment of the logic of relatives, as I contend, will dispense altogether with class names and only use ... verbs.

This was probably effected because propositions could now be simplified to a variable signifying the relation and a quantifying symbol. In other words this emphasis on the relation rather than the class lent itself to his newly developed quantificational theory of logic.

7.2.4. A Note on the Grassmanns. The main influences on Peirce's logic apart from his father Benjamin, Boole and De Morgan, seem to have come from Mitchell, his student and to a lesser extent, the Grassmann brothers. Herman Grassmann (1809–1877) was a schoolteacher who in 1844 represented geometric objects together with their combinations in an algebra that was capable of commutativity and distributivity. His brother Robert (1815–1901) also established a system of logic where objects of thought could be composed as sums of 'pegs' 'e'. This Boolean type algebra seems to have been developed totally without the knowledge of the work of Boole or Jevons (Grattan-Guinness 2000, ch. 4). Whilst Schröder was initially influenced by the Grassmanns before he discovered Boole, Peirce does not seem to be greatly influenced

---

[12]See [**83**, p. 548]: the concept "class" is formed by observing and comparting class-concepts and other objects and also his theory of abstraction.

[13]See [**84**, p. 290].

by them and commented that Robert Grassmann's treatment "presents inequalities of strength; and most of his results had been anticipated." He also claimed that Hugh MacColl's papers on a "Calculus of Equivalent Statements" [118] were nothing but Boolean algebra, with Jevons's addition and a sign of inclusion.[14] Other influences to be found in a completely different area came from the work of A. B. Kempe and the topology of Johann Listing, as we shall see in Section 7.4.2.

7.3. Quantification and Iconicity

7.3.1. The Development of Quantification Theory. Let us now review the development of quantification in Peirce's logic. As can be seen from *Logic of Relatives*, Peirce described universal propositions in a Boolean way: h,(1-a)=0 or ALL HORSES ARE ANIMALS or using illation which expressed inclusions between simple relatives: h≺a. For particular propositions he used h,b≻0 to mean SOME HORSE IS BLACK where h≻b but it is not true that b≻h. This can also be expressed using Peirce's quantifying operation of involution as in $0^{h,b} = 0$ (or in other words members of the class h,b exist). The inspiration behind this quantification was Boole. Peirce at this stage in *Logic of Relatives* wanted to continue Boole's equational program of expressing all propositions as equations and this provided a reason for quantifying his logic of relatives. Later he confirmed illation ≺ as the fundamental operation rather than = and this took him away from equational logic, thus a new method of quantification was necessary.

The development of quantifiers by Peirce first started from his algebraic inclination to use linear combinations, i.e., to express relative terms as infinite sums by attaching a logical coefficient that took the value 1 or 0 to indicate existence or nonexistence respectively. Using the symbol Σ as a shorthand notation for an infinite sum, existence for a particular term can be denoted by a simple inequality:

$$\Sigma_i \Sigma_j l_{ij}(\text{I:J}) > 0$$

means that there exists i and j such that i LOVES j. By discarding the inequality and the ordered pair (I:J), it is clear that the symbols and the logical coefficients alone are sufficient to indicate existential quantification, i.e., $\Sigma_i \Sigma_j l_{ij}$; meaning THERE EXISTS i AND j SUCH THAT i LOVES j.

---

[14]See [**84**, p. 126].

The breakthrough occurred when he took from Mitchell the idea of having one symbol for universal quantification and a different symbol for existential quantification, but the choice of Σ and Π as quantifier symbols used as free-standing operators rather than using subscripts attached to terms to represent quantification, was entirely his original conception. Furthermore, as outlined above, Peirce soon discovered that his formulation could be simplified because with the quantifier symbols there is no need for a logical coefficient and l_{ij} soon came to denote the ordered pair and the relation between them, rather than a coefficient which takes the values 0 or 1. Later a further simplification occurred when the inequality > 0 was also omitted and quantification was then expressed in terms of a quantifier symbol and a logical term indicating the ordered pair and the relation.

Why were these particular quantifier symbols Σ and Π chosen? Firstly existential quantification had come to symbolize for Peirce, an existent term in a linear combination of individual terms as per his definition of a relative term. This occurred as early as 1870 in *Logic of Relatives*. If such an individual term did not appear in such a combination then it did not exist. The symbol for an infinite sum then became linked with existential quantification and the next step was to introduce Boolean coefficients that ensured existence or non-existence in terms of an inequality.

On the other hand universal quantification, I suggest, could have been influenced by Mitchell[15], in the sense that Mitchell found a general expression for universal propositions as products of the standard De Morgan propositions. I would argue that this gave Peirce the inspiration to use the infinite product symbol Π to denote universal quantification. So the origins of his theory of quantification first started in *Logic of Relatives*, when he defined relative terms as linear combinations of individuals. This progressed to a definition of a relative term as an aggregate of ordered pairs by 1882. The quantifying symbol was then treated as an operator obeying certain rules. By 1885, he was expressing quantification without recourse to numerical coefficients. Propositions were now composed of a Boolean expression referring to an individual and a quantifying part specifying the individual.

It is interesting to speculate on whether Peirce knew that with his quantificational theory he had a potentially infinite language, and if so, was he then aware of any problems inherent in the system. It seems likely that the answer to the first question is

---

[15]See [**128**].

yes, but it has not been possible to discover whether he was then able to comprehend or solve any of the concepts and difficulties associated with such an infinite language.

We should now ask the question, why after the success of his algebraic logic should Peirce now abandon it in favor of a quantificational theory of relations? This theory developed after 1882 was highly rated by Peirce, over his previous algebraic logic primarily by scoring in terms of deductive convenience. Whereas the earlier algebraic logic was difficult to use especially when plural relatives between three or more objects were involved, quantificational logic with its two-part formulation of quantifier and relation had the advantages of simplicity and convenience. The quantificational theory also scored highly in terms of expressive power and generality of method. As Merrill states:[16]

> the [algebraic logic of relations] contains too many ways of doing
> the same thing, while the [quantificational logic of relations] has
> one unified method.

In this respect it is also superior in terms of analytical depth and more crucially to Peirce, also appeared to be more fundamental in logical terms.

7.3.2. The Influence of Peirce's Quantificational Theory. It was Peirce's work on first order predicate logic in his 1885 paper on quantificational logic, as extended by Schröder in [**186**] that was a primary influence on Löwenheim and Skolem. The route of influence can be traced from Peirce to Schröder to Korselt to Löwenheim and Skolem as reflected by their notations and methods.[17] Hiz [**94**] shows that Peirce also influenced logic in Poland and in particular the work of Wajsberg and Lukasiewicz who looked at the axioms Peirce had formulated in his quantificational logic of 1885.

In his quantificational logic, Peirce also introduced logical values. These are values that may be taken by propositions. For him these values are the constants 0 and 1 and all values in between. In the binary system of logic **f** and **v** are used for 0 and 1 and can be used as propositional terms in logical formulæ. His treatment of hypothetical propositions, taking 'if p then q' to be true (or =**v**) except when p is true and q is false,

---

[16]See [**125**, p. 171].
[17]See [**16**, p. 173].

is now generally accepted. Peirce's logical values were generally used by the Warsaw logicians, Lukasiewicz, Leśniewski, Tarski and Wajsberg.

7.3.3. The Iconic Notation for the Sixteen Binary Connectives.

By 1880, in "On the Algebra of Logic," Peirce had defined the table of sixteen forms of the logical binary connectives for the first time. When considering the logic of two propositions X and Y, there are sixteen possible relations between these propositions such as X AND Y, X OR Y etc. These relations are often called the sixteen binary connectives. First considered by De Morgan, it can be seen that using three individuals and two relations, sixteen possible propositions can be formed, with the overstrike indicating the converse of the relation, such as $(\overline{A:B})(B:C)$.[18]

Peirce's icons were first introduced in 1902 in a section called 'The Simplest Mathematics' which is one section of "Minute Logic" (MS 429, CP 4.227-4.323) and take the shape of an X-frame. The binary connectives are shown below in sixteen columns, each containing four entries. These entries indicate the quadrants that are closed (E) or left open(T) in Peirce's X-frame, with the entries in rows one, two three and four applied, respectively, to the top, left, right and bottom quadrants of the frame. Peirce's icons appear below. He later modified these icons to a more rounded cursive script. In particular he used the sign V for ⋈. This brings to mind Ladd-Franklin's symbol for her logical operation introduced twenty years previously in [**110**] that we covered in Chapter 5. In fact this sign is used for substantially the same purpose, and moreover, in manuscript 431A: 56 he wrote:[19]

> It ought to be remembered that it was Mrs. Franklin who first pro-
> posed to put the same character into four positions in order to repre-
> sent the relationship between logical copulas, and that it was part of
> her proposal that when the relationship signified was symmetrical,
> the sign should have a left and right symmetry.

It should be noted however that Ladd-Franklin got stuck on only half of the connectives and did not develop the full system.

---

[18]See section 5.2.3 above.
[19]See [**203**, p.336].

In this system as shown in Table 7.1 $A \boxtimes B$ represents 'A is not true but B is true' and $A \boxtimes B$ represents IF A IS TRUE, SO IS B. Peirce included $A \boxtimes B$ and $A \times B$ in his list, although he called the former EVERYTHING IS IMPOSSIBLE (ABSURD)' and the latter MEANINGLESS.

1	2	3	4	5	6	7	8	9	10	11	12	13	14	15	16
F	F	F	F	T	T	T	T	F	F	F	F	T	T	T	T
F	F	F	T	F	T	F	F	T	T	F	T	F	T	T	T
F	F	T	F	F	F	T	F	T	F	T	T	T	F	T	T
F	T	F	F	F	F	F	T	F	T	T	T	T	T	F	T
⊠	⊠	⊠	⊠	⊠	⊠	⊠	⊠	⊠	⊠	⊠	⊠	⊠	⊠	⊠	✕

TABLE 7.1. Sixteen Iconic Binary Connectives

It seems that by inventing the iconic notation, after having received the initial inspiration from Ladd-Franklin, Peirce proved yet again to be a remarkable innovator regarding future applications to crystallography and group theory. Zellweger claims:[20]

> The logic of propositions is a fundamental part of symbolic logic. If one gives central emphasis to the role of symmetry, when great care is put on shape designing what it takes to construct a special set of sixteen iconic signs, then it is possible to bring to the logic of propositions an approach that not only simplifies and consolidates.
>
> This approach, with its emphasis on symmetry, also receives major assistance from the algebra of abstract groups ... It has practical implications for digital design, mirror logic, and optical computers.

Clark [28] gives a detailed description of how Peirce constructed his notation and how he used it. Current research on the icons that Peirce devised for these sixteen connectives has shown that this important work was neglected, first because major sections were omitted from the publication of his logical works (Collected Papers) in the 1930s, and second because Peirce's signs were replaced by other signs of the editors' choosing.[21]

---

[20]See [203, p.76].
[21]See [28, p. 305].

7.4. The Relation between Logic and Mathematics

7.4.1. Overlapping Disciplines. George Boole was a pioneer in establishing the close relationship between logic and mathematics. Boole believed logic and mathematics to be separate branches of a universal language but used the forms and methods of mathematics to formulate an inferential calculus of logic. He claimed that the ultimate laws of Logic are mathematical in their form. In this sense, he was applying mathematical techniques to logic. The position of De Morgan regarding the relation between logic and mathematics was very similar to that of Boole. He considered logic and mathematics as separate, but used algebraic analogies e.g. in his use of symbols to represent quantification, and in the opposing operations of addition and subtraction to consider logical inverses and complements. Algebraic considerations also led him to abandon the traditional copula of identity 'is' and replace it with a more general relation that was both transitive and convertible such that all inference in logic could now be represented by composition of such relations.

Although logic could be defined broadly as the inquiry into truth and falsehood and narrowly as the investigation of the Aristotelian syllogism, De Morgan favored Kant's definition as 'the science of the necessary laws of thought', a definition that echoed both Benjamin Peirce's definition of mathematics and Boole's *Laws of Thought*. De Morgan was clear that mathematics is not part of logic, or vice versa; the main distinction being that logic deals with the pure form of thought without any considerations of matter, but he did feel that there are two parts to logic, mathematical and metaphysical. Like Boole, he felt that logic and human reasoning could be symbolized in the forms of algebra. This he left this largely to Boole, contenting himself with reforming the traditional Aristotelian syllogism and developing a theory of relations.

What was Charles Peirce's own position on the relation between logic and mathematics? By 1902 he was aware of the logicist programme of Dedekind[22] but disliked it. Even his former student Christine Ladd-Franklin in her article with E. V. Huntington for vol. 9 of The *Encyclopaedia Americana* [**100**] mentions Russell as having announced 'the surprising thesis that logic and mathematics are in reality the same science; that pure mathematics requires no material beyond that which is furnished by the necessary presuppositions of any logical thought; and that formal logic, if it is to

---

[22]"The philosopher mathematician, Dr. Richard Dedekind, holds mathematics to be a branch of logic." [**85**, p. 239].

be distinguished as a separate science at all, is simply the elementary, or earlier, part of mathematics'.

However, it is clear that Peirce was not a logicist, i.e., he did not consider that logic is the foundation of mathematics. For him, mathematics and logic had clear distinctions. In order to see this, one must look at his philosophy and his classification of the sciences where according to his classification; mathematics is at the foundation of the sciences. Logic comes later. That is not to say that logic can be reduced to mathematics, rather that logic depends essentially on the results of mathematics, i.e., that it uses data or concepts from mathematics. Logic to Peirce was the study of the methods by which inferences could be drawn. He was interested in working out a philosophically satisfying system of logic, one that gives the most revealing account of basic inferential steps – the search for analytical depth. His conception of logic as a science often stressed this objective over mere manipulative ease. This explains the three successive logical systems developed over his career: algebraic, quantificational and graphical. Some commentators such as Dipert have even claimed that Peirce espoused a kind of reverse logicism.[23] I would agree with this in so far as Peirce stated several times that logic depends on mathematics, but this is not the same as holding that logic is part of mathematics.

It is also the case that Peirce's work on the foundations of mathematics e.g. his 1881 paper on the axiomatization of arithmetic was considered by him as mathematics rather than logic. Commentators such as Stephen Levy have attempted to show that Peirce was reluctant to acknowledge the role of logic in mathematical reasoning because of his clear distinction between the two disciplines of logic and mathematics.[24] However, the examples that Levy cites are all foundational, and so there would have been no conflict on the part of Peirce to see such work as mathematics and not logic. In fact, Levy tries to show that Peirce does claim that mathematics depends on logic, but unfortunately the very concepts that Levy chooses to support his case are Peirce's discussions of set theory, the very concepts that Peirce argued are properly in the realm of logic not mathematics. It was, as suggested by Levy, Peirce's love and respect for his father Benjamin Peirce, a distinguished mathematician, that led him to reject the mutual dependencies of logic and mathematics. However I suggest there is evidence to suggest that it was more a case of Charles influencing his father's position on this matter.

---

[23]See [49, p. 57].
[24]See [114].

Benjamin discussed the relationship between logic and algebra in *Linear Associative Algebra*. 'Mathematics', he said, 'belongs to every enquiry, moral as well as physical. Even the laws of logic, by which it is rigidly bound, could not be deduced without its aid.'[25] As it is described here there is a kind of reciprocity between the two: logic is more fundamental in a sense, but it is mathematics which transmutes logical form to fit natural language. Mathematics is an arbiter with respect to the world at large. His father's unwillingness to assign either logic or algebra an absolute primacy over the other was according to Charles's later account, the result of his dissuading his father from the view then held by Dedekind that mathematics is a branch of logic. "I argued strenuously against it,' Charles says, 'and thus [Benjamin] consented to take the middle ground of his definition'[26], which suggests again that neither logic nor mathematics was assigned primacy.

As Grattan-Guinness has pointed out, although mathematics (and in particular algebra) was applied to algebraic logic, Peirce also saw a task for logic in analyzing mathematics. For him, as with as his father, it was a case of mathematics drawing conclusions, logic theorizing about such conclusions. He wrote 'My own studies in the subject [of algebra] have been logical not mathematical, being directed towards the essential elements of the algebra, not towards the solution of problems' (Peirce 1882, postscript). And again on 2 April 1908, he wrote in an unpublished letter to C. J. Keyser of Columbia University cited on page 12 of **[98]**, highlighting the difference between mathematics and logic by contrasting mathematicians and logicians:

> As for the difference between the mathematician and the logician - and the two kinds of thought have nothing in common except that both are exact - it is that the mathematician seeks the solution of a problem and has but a subsidiary interest in anything else, while the logician, not caring a snap what the solution may be, desires to analyze the form of the process by which it is reached, in order to get a general theory of the form of intellectual procedure.

Elsewhere he wrote 'the greater number of distinct logical steps, into which the algebra breaks up an inference will constitute for [the logician] a superiority of it over another

---

[25]See [**137**, p. 97].

[26]MS 78,4.

which moves more swiftly to its conclusions. He demands that the algebra shall analyze a reasoning into its last elementary steps.'[27] It should be understood that here Peirce uses the word 'algebra' for 'algebra of logic'. As for 'not caring a snap what the solution may be', although not directly concerned with finding applications for his algebraic logic, it was obviously important that solutions to problems would eventually be found using his logical methods and correct ones at that. It is also clear that Peirce saw logic as a means of revealing new mathematical methods. One example of this is evident in [154] when he considers the syllogism:

(1) Every Texan kills a Texan
(2) No Texan is killed by more than one Texan
(3) Hence every Texan is killed by a Texan[28]

This form of the syllogism due to De Morgan is called the syllogism of transposed quantity. It prompted Peirce to discover that a multitude must be finite if no one-to-one correspondence could be found between the multitude and any proper subclass, thus specifying the difference between finite and infinite multitudes well before Dedekind in 1888.[29] However although inspired by his logical studies to investigate the concepts of continuity and infinity, he would have regarded these areas as logical rather than mathematical.

His work on the logic of relatives also led to important contributions to linear algebra and the theory of matrices. By expressing his elementary relatives as algebras with n individuals he could produce an algebra with n units: using a Universe with two individuals $\{u, v\}$ we have as individual relatives: $(u, u) = c$, COLLEAGUE OF; $(u, v) = t$, TEACHER OF; $(v, u) = p$, PUPIL OF and $(v, v) = s$ SCHOOLMATE OF. With multiplication obtained by composition of relations we have

$$(u, u)(u, u) = u, \qquad (u, v)(v, u) = u$$

---

[27]See [84, p. 239].

[28]In another version of this syllogism, Peirce drew from Balzac's introduction to the *Physiologie du mariage*, and recast the syllogism in terms of the seduction of French women. This is very interesting in the light of his second manage to Juliette Portelai (or Froissy). See [129, p. 259].

[29]Dauben [31] asserts that Peirce had claimed that Dedekind's famous monograph, *Was Sind und was Sollen die Zahlen*, had been influenced by his own work, because Peirce had sent a copy of his 1881 paper to Dedekind. However Peirce was wrong as the the main part of Dedekind's 1881 monograph was drafted in 1872.

and
$$(u, v)(u, v) = 0.$$

These are the same rules of multiplication as can be observed in *Linear Associative Algebra* and in this way, Peirce was able to represent all linear associative algebras in the form of his relative terms. The above logical terms can be given as a matrix as shown in Table 7.2.

	c	t	p	s
c	c	t	0	0
t	0	0	c	t
p	p	s	0	0
s	0	0	p	s

TABLE 7.2. 4×4 Multiplication Matrix

This gives an algebra of dimension n^2. Peirce recognized the similarity of his work with Cayley's work on matrices and Sylvester's work on nonions. In fact by the spring of 1882 he had realized that his algebra which consisted of linear combinations of n individuals was 'mathematically identical to the algebra of all n×n matrices over the given field'[30]

In a series of letters kept in the Sylvester Papers in St. John's College, Cambridge (kindly provided by A. Crilly), Peirce explains to Sylvester the relationship between his relative terms and Sylvester's matrix algebra. The first letter dated January 4 1882 considers how the notation for his relative terms could apply to matrices:

1882 Jan[uary] 4.

My dear Professor

When I saw you the other day I had not received yours of the 30th ultimo. I think it would be useful to write $(a)_{ij}$ to denote the quantity which stands in the ith column and j^{th} line of the matrix denoted by $\underline{a}$. At any rate, this notation will enable me to express what I want to say now. We may conceive matrices as subject to two

[30]See [**102**, p. 198].

series of operations, the internal and the external. The internal combinations are defined by this formula which defines a combination denoted by φ of the two matrices a and b.

$$(\varphi(a,b))_{ij} = \varphi((a)_{ij}, (b)_{ij})$$

Two more letters were written on the 5th and 6th January. The first of these has been published in [**134**, pp. 205–207]). Peirce presses his claim that the multiplication of his relative terms is exactly that of matrix multiplication. He is thinking here of his addendum to Benjamin Peirce's *Linear Associative Algebra* published in 1880 in which he outlined how linear associative algebras could be represented by his relative terms written in a matrix notation.[31] The Jan 5th letter starts:

21 Read St.
1882 Jan[uary] 5

My dear Sir

The precise relationship of your algebra of matrices to my algebra of relatives is this. Every relative term, according to me, consists of a sum of individual relatives each affected with a numerical coefficient. When the relative is a *dual* relative, the individuals naturally arrange themselves (& I always arrange them) in a matrix. Hence their coefficients may be arranged in a matrix. Now, the matrix of coefficients of what I call the product of two relatives is precisely what you call the product of the two matrices that are formed by the coefficients of the factors.

A few lines later Peirce claims:

It, thus appears to me just to say that the two algebras are identical, except that mine also extends to triples & other relatives which transcend two dimensions.

This was presumably enough to alarm Sylvester and prompted Peirce to write again on the following day in an unrepentant style:

---

[31] See pp. 190–191 above.

What I lay claim to is the mode of multiplication by which as it appears to me this system of algebra is characterized. *This* claim I am quite sure that your own sense of justice will compel you sooner or later to acknowledge. Since you do not acknowledge it now, I shall avail myself of your recommendation to go into print with it. I have no doubt that your discoveries will give the algebra all the notice which I have always thought it merited and therefore I hope my new statement of its principles will be timely. I cannot see why I should wait until after the termination of your lectures before appearing with this, in which I have no intention of doing more than explaining my own system & of saying that so far as I am informed it appears to be substantially identical with your new algebra, & that it ought to be, for the reason that mine embraces every associative algebra, together with a large class perhaps all of these which are not entirely associative. I am sorry you seem to be vexed with me.
Yours very faithfully
C. S. Peirce.

The final letter of the series written on 5 March 1882 refers to the proof contained in Peirce's 1880 Addendum paper although he does not mention this paper by name:

I have a purely algebraical proof that any associative algebra of order n can be represented by a matrix of order $n + 1$ having one row of zeros, together with a rule for instantaneously writing down such a matrix.

Grattan-Guinness [71] has also noticed that Peirce's expansion theorems possessed the same algebraic structure as Cayley's formula for an element in the product of matrices. Both were also isomorphic with another important algebra of the time: the scalar product of two vectors. Because Peirce's algebraic form of expressing relations can be represented by matrices since they both have the same rule of multiplication Lenzen [113, p. 245] claims that the concepts introduced by him made possible the linear representation of a matrix, as each element of the matrix could be expressed as a linear combination of the product of the element and a unit matrix. However the unit matrix was then easily expressible as an elementary relative in Peirce's algebraic logic. Iliff also claims[32] it was Peirce's thorough understanding of matrix techniques and in

---

[32]See [102, p. 201].

particular matrix multiplication that led to the connection with quantification so that 'Peirce's discovery of the quantifiers was based on his expertise in sophisticated techniques of abstract algebra; it was not merely a simple generalization of the Boolean sum and product. Thus an application of mathematics led to an advance in logic'.

Influenced by his father Benjamin Peirce's famous definition of mathematics as 'the science which draws necessary conclusions', he in turn defined logic as 'the science of drawing necessary conclusions'. It is also clear that Boole, De Morgan and even initially Peirce hoped to apply their results in particular to the field of probability.

7.4.2. A Topological Turn, 1889-1898. Peirce took a new path by developing firstly in 1889 and then after 1897 a graphical form of logic called the existential graphs. This was to be the 'logic of the future' and Peirce did very little algebraic logic work after this time. Initially inspired by Alfred Bray Kempe, Peirce moved from his early form the entitative graphs to his final form - the existential graphs. Kempe's *Memoir on the Theory of Mathematical Form* of 1886 had a profound and lasting effect on him. Kempe introduced a graphical notation of spots and lines (bonds) modeled after the chemical tree diagrams which showed the constitution of compounds. The spots in Kempe's system represent 'units': the entities in terms of which the mind, in the process of reasoning, deals with the subject matter of thought. 'These units come under consideration in a variety of garbs - as material objects, intervals or periods of time, processes of thought, points, lines, statements, relationships, arrangements, algebraical expressions, operators, operations etc., etc.[33]

One of the major shortcomings of Boolean algebraic logic was its difficulty in expressing mixed hypothetical and categorical statements. By using a multidimensional logic, i.e., a universe for predicate terms and a universe for propositions, Mitchell solved one of the major inadequacies of most Boolean theories. Peirce greatly appreciated these multi-dimensional logical universes and praised Mitchell's work in this area as one of his most important contributions to exact logic. By considering these universes, he was able to link his later development of the existential graphs with its topological models, i.e., the Sheet of Assertion as a surface representing the universe of actual existent fact, and a book of such sheets models other possible existential universes where the cuts are means of passing from one universe to another. This added the important concept of modality to his logical system, and anticipated the 'possible

---

[33]See [**105**, p. 2].

worlds' semantics for modal logic that has proven so powerful and fruitful in the study of modal logic.[34]

Peirce was able to do this by using the diagrammatic tools of chemistry to graph logical relations not only between relational terms and classes but also between equations. As such, his logical graphs became a new means to geometrically represent logical implications among propositions, in a more sophisticated and complex way than could be done by Venn diagrams. The existential graphs have tended to be ignored until recently when Roberts [171] supplied a much needed analysis of the graphs that were unavailable except in the *Collected Papers* edition. Apart from the factor of availability, one reason for this ignorance could be that the graphs were intended to facilitate the analysis of logical structure – a point of key importance to Peirce – rather than as a calculus to draw inferences. As such the existential graphs need a certain amount of practice for the reader to become familiar with the necessary techniques required. It is a tool for analysis rather than a quick and easy calculus. Recently modern-day Peirce scholars such as Burch and Brunning have attempted to show that Peirce's Reducibility Theorem holds, i.e., that all *n*-adic relations can be reduced to dyadic and triadic relations, but the main problem that they face is that their proofs are diagrammatic and have not yet been shown to be algebraically justified.

As we have seen in Chapter 5, Peirce progressed from a quantificational system using numerical identities and inequalities to one in which the quantifiers are attached to logical terms which represent objects and relations, i.e., propositions. It was this step which was vital to his system of logical graphs. In a basic system illustrated below the lines represent individuals - something or someone. These are contained in an unpublished letter to Mitchell dated Dec 21,1882, MS 294, cited in [171].

FIGURE 7.1.   Someone is both a benefactor of and loved by something. (Conjunction) i.e. a benefactor of a lover of itself. Algebraically: $\Sigma_x \Sigma_y b_{(xy)} l_{(xy)}$

Propositions with a combination of quantifiers are dealt with thus:

---

[34]See [205, pp. 241–256].

FIGURE 7.2.   Someone is both a benefactor and a lover of something.(Conjunction) $\Sigma_x\Sigma_y b_{(xy)}l_{(yx)}$

FIGURE 7.3.   Someone is a lover of itself. Algebraically: $\Sigma_x l_{xx}$

FIGURE 7.4.   Everything is a lover of itself. (Universal quantification) $\Pi_x l_{(xx)}$

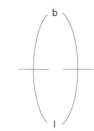

FIGURE 7.5. Everything is either a benefactor of a lover of every-thing. Note that universal quantification changes conjunction into disjunction. Algebraically: $\Pi_x\Pi_y b_{(xy)} + l_{(xy)}$

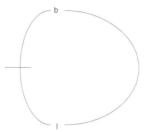

FIGURE 7.6. Everything is both a benefactor and a lover of some-thing.(Conjunction) Algebraically: $\Pi_x\Sigma_y b_{(xy)} l_{(xy)}$

Note that Peirce does not express the proposition SOMEONE IS EITHER A BENEFACTOR OR A LOVER OF SOMETHING.

In a manuscript dated January 15,1889, entitled 'Notes on Kempe's Paper on Mathematical Forms', the idea of representing individuals by the lines of the diagrams rather than by the spots (as Kempe did) occurred to Peirce:

> These ideas of Kempe simplified and combined with mine on the algebra of logic should give some general method in mathematics.

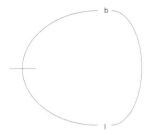

FIGURE 7.7.   Everything is either a benefactor or a lover of something.(Disjunction) Algebraically: $\Pi_x\Sigma_y b_{(xy)} + l_{(xy)}$

The next stage was the Entitative Graphs – a system of representing the logic of classes or propositions as logical graphs, which was first introduced in [**160**]. Propositions or predicates are placed in linked circles. It is thus possible to express non-relative propositions. Encircling the proposition represents negation:

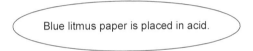

FIGURE 7.8.   Negation

means IT IS FALSE THAT BLUE LITMUS PAPER IS PLACED IN ACID. Disjunction is represented by adjacent propositions:

Blue litmus paper is placed in acid.

Blue litmus paper will turn red

FIGURE 7.9.   Disjunction

means EITHER BLUE LITMUS PAPER IS PLACED IN ACID OR THE BLUE LITMUS PAPER WILL TURN RED. Implication is represented by encircling the antecedant proposition:

Conjunction is represented by using three circles:

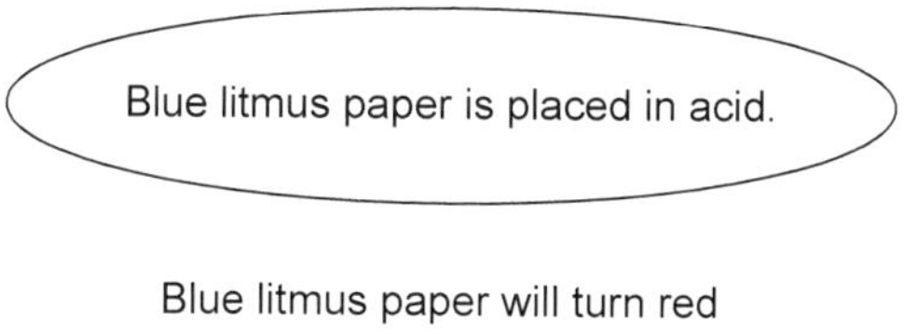
Blue litmus paper is placed in acid.
Blue litmus paper will turn red

FIGURE 7.10. Implication

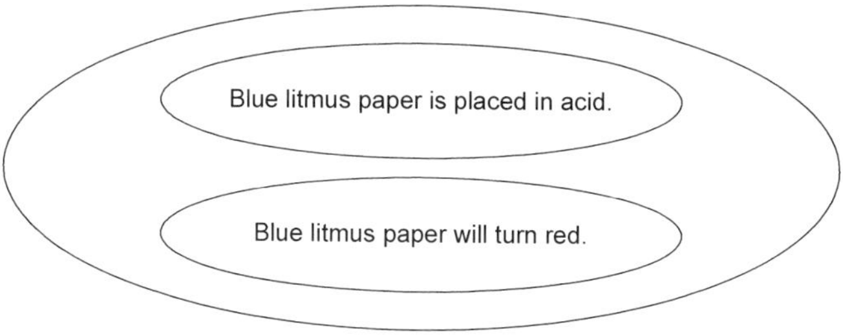
Blue litmus paper is placed in acid.
Blue litmus paper will turn red.

FIGURE 7.11. Conjunction

To encompass quantification, Peirce used the device of a line to signify either universal or existential quantification following the rule: If the line's least enclosed part is enclosed by an odd number of circles, then this signifies SOME, otherwise the signification is EVERY. Consider the following examples:

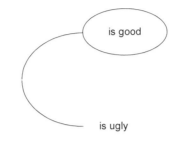

FIGURE 7.12.   Everything good is ugly

Here the least enclosed end of the line is not encircled and we have the conditional form so we have the meaning EVERYTHING GOOD IS UGLY.

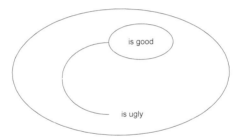

FIGURE 7.13.   Something that is good is not ugly

Here the least enclosed end of the line is encircled once and we have the conjunctive form so we have the meaning SOMETHING THAT IS GOOD IS NOT UGLY.

Here we have the line enclosed by one circle to give SOME and the negative conditional form which represents IF SOMETHING IS GOOD THEN IT IS FALSE THAT IT IS UGLY or NOTHING GOOD IS UGLY.

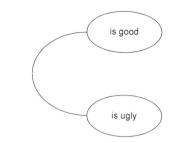

FIGURE 7.14.   Nothing good is ugly

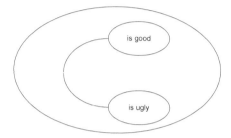

FIGURE 7.15.   Something good is ugly

This represents the conjunctive form SOMETHING THAT IS GOOD IS ALSO UGLY or SOME-
THING GOOD IS UGLY.

Peirce then introduced a new system of graphical diagrams called the Existential
Graphs because he believed it to be simpler and easier to use. In this system, the interpretation of Figure 7.8 remains unchanged but the interpretations for Figures 7.9
and 7.11 are interchanged so that Figure 7.9 now represents conjunction and Figure
7.11 represents disjunction. The representation of conjunction is therefore more intu-
itive since to assert a proposition is to write it down; so as in Figure 7.9 to have two
propositions written together, is to assert both.

Implication or IF ... THEN statements are expressed as

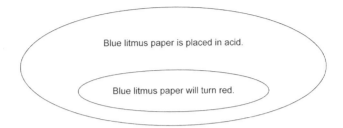

FIGURE 7.16.   Implication

rather than Figure 7.10.

Quantification is also expressed by using a line connecting predicates, but using the reverse interpretation, so that we have the reverse of the rule for entitative graphs: If the line's least enclosed part is enclosed by an odd number of circles, then this signifies EVERY, otherwise the signification is SOME. The new representations of the conditionals in Figures 7.13 and 7.12 are:

FIGURE 7.17.   Everything good is ugly

This is in the form of a conditional, and the least enclosed part of the line is enclosed once, i.e., oddly so we have the universal quantifier EVERYTHING IF IT IS GOOD IS UGLY or EVERYTHING GOOD IS UGLY.

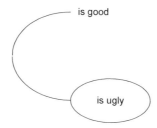

FIGURE 7.18. Something good is not ugly

This is expressed in the form of a conjunction so that we have SOMETHING IS GOOD AND IS NOT UGLY or SOMETHING GOOD IS NOT UGLY.

7.4.3. Signs and Triads. Peirce devoted years of his life to developing powerful algebraic systems of logic. Why, then, did he not devote his later logical efforts to the improvement of these algebraic logics, rather than to the formulation of a new graphical system of logic so radically different in symbolism from those of his earlier logics? As a system of logic the existential graphs consist of the Alpha system corresponding to nonrelative logic, the Beta system to first intentional logic of relatives, and the Gamma system to second intentional logic of relatives. The answer is that he considered the existential graphs the best way of representing the separate inferences necessary to describe a system of logic in the most detailed and clear way. Also the fact that the existential graphs could describe such a system iconically also had its benefits for Peirce the semiotician.

It is not in fact so surprising that he took this turn when we consider firstly that Peirce claimed that he 'thought in diagrams' by default, as it were. Let us review his philosophy. In his general theory of signs, he considered that all thought is in signs. Cognition is a three termed or triadic process. Operations upon symbols, including substitution are present in all logical thought. Logic calls for a mixture of icons, indices and tokens (symbols). Even though central importance belongs to the logic of relations and even though logic is not the whole of semiotic, logic in full is logic that is wholly semiotic. A second consideration to note is that Peirce devoted his life to the search for the definitive logical system. First relational algebraic logic was developed, to be followed by his quantificational theory of logic. The reason Peirce made the effort of developing the graphs was that he felt that they would more effectively perform the function of logic. The development of a thorough understanding of mathematical reasoning was, Peirce believed, the purpose for which his logical algebras were designed, but he felt that his new system of existential graphs was far more effective in that respect.

Benjamin Peirce's definition that mathematics is 'the science which draws necessary conclusions' was taken by Charles to mean that mathematical reasoning is deductive reasoning and is therefore an integral part of reality itself. To understand reality an analysis of the deductive process 'by breaking up inferences into the greatest number of steps, and exhibiting them under the most general categories possible' is needed;[35] for this purpose he came to prefer the existential graphs to the algebras of logic. According to Zeman [**204**], the main reason is that such a system is better able to express continuity.

In algebra, two occurrences of an identical symbol, e.g., the variable a, represent the same object. However this is represented by a *line of identity* in the existential graphs. The individual is far better represented, Peirce felt, by a continuous line than by a number of discrete occurrences of an individual variable since as Peirce puts it 'the line of identity which may be substituted for the selectives very explicitly represents Identity to belong to the genus Continuity.'[36] As well as the line of identity, in the Alpha and Beta systems, *cuts* which are the negation signs of these systems are discontinuities on the sheet of assertion on which the graphs are drawn and correspond to discontinuities in reality; the sheet of assertion being a two dimensional space or surface which represents the actually existing universe. This emphasis upon continuity is more readily understood when we come to realize that in his later years Peirce was fascinated with topology. Furthermore, according to Murphey [**129**] the model on which Peirce based his metaphysics was the topology of Listing, and his work in mathematics had led him to the conclusion that topology is the mathematics of pure continua. In his graphs, Peirce used topology to model logic as he had once used algebra. As Zeman states:[37]

> Thus the graphical systems may be considered to be kind of a "topology of logic," and modelling their symbol structure upon topology

---

[35]See [**85**, p. 229].
[36]See [**84**, p. 561 f.].
[37]See [**204**, p. 150].

and its continua, they are better able, for Peirce, to represent reality which is, after all, continuity'.

Another reason for favoring a graphical system rather than an algebraic system of logic is to be found in one of the major shortcomings of Boolean algebraic logic, i.e., its difficulty in expressing mixed propositional and categorical statements. By using a multi-dimensional logic, i.e., a universe for predicate terms and a universe for propositions, Mitchell solved this inadequacy. Peirce greatly appreciated the concept of multi-dimensional logical universes and praised Mitchell's work in this area as one of his most important contributions to exact logic. By considering these universes, he was able to link his later development of the existential graphs with its topological models, i.e., the Sheet of Assertion as a surface representing the universe of actual existent fact, and a book of such sheets model other possible existential universes where the cuts are means of passing from one universe to another. This added the important concept of modality to his logical system, and anticipated the 'possible worlds' semantics for modal logic that has proven 'so powerful and fruitful in the study of modal logic', according to Zeman [205].

An alternative answer lies in Peirce's logical beliefs, in particular his Reducibility Theorem or theory of categories, which claimed that all thought, is divided into three irreducible categories: monads, dyads and triads. These categories suggested the existential graphs. Brunning shows that Peirce's algebras failed to provide a compelling explicit demonstration that the third category of triads was necessary, whereas the existential graphs with their extra dimension does so.[38] Using the existential graphs, modern-day Peirce scholars such as Robert Burch have attempted to show that Peirce's Reducibility Theorem is valid, but the main problem that they face is that their proofs are diagrammatic and have not yet been shown to be algebraically justified.[39] Existential graphs, as well as having a direct translation to language, are also isomorphic to the discourse representation structures that were independently developed over eighty years later. These structures use variables to represent discourse referents which correspond to existentially quantified variables. Boxes represent contexts and arrows represent implication. Kamp's discourse representation theory in [104] attempts to express the constraints involved when new contexts are introduced by negations, modalities or time. John Sowa claims in [191] that these structures are isomorphic to Peirce's existential graphs. Sowa's *Conceptual Structures: Information Processing in Mind*

---

[38]See [20].
[39]See [21].

and Machine of 1984 uses conceptual graphs which are an extension of Peirce's existential graphs. The existential graphs also form the logical foundation for conceptual graphs, which combine Peirce's logic with research on semantic networks in artificial intelligence and computational linguistics.

7.5. The Influence of Peirce on Later Logicians and Mathematicians

Pierce's hope that his algebraic logic would open up new methods in logic, had to some extent been already justified by the expressive power of the logic in dealing with categorical and hypothetical propositions – always a weakness of Boolean logic. Moreover, he hoped that his logic would also lead to new developments in mathematical methods because the analytical depth of the logic would reveal new methods in mathematics. He also cited applications of his logic to linear algebras, quaternions, vector geometry and even political economy. However the algebraic logic of 1870 was superseded by his quantificational logic of 1885. What Peirce did not do was to provide a formal notion of formula or of proof in his quantifier logic, but instead he concentrated on the semantic interpretation of the system. This fact is not surprising given the cursory proofs of many of his logical axioms. For Peirce, the primary concept was that of analytical structure and expressive power of a logical system.

Schröder developed Peirce's relational logic as a foundation for mathematics. Volker Peckhaus has shown [**136**] that this influence led to a deep change in Peirce's conception of the role of logic in founding mathematics. His algebra and logic of relatives became the basis of a scientific universal language. In this Schröder differed fundamentally from Peirce. He did not deal with the application of mathematical methods to the analysis of logic, but rather following the logistic programme, with the description and analysis of mathematics by the means of logic. Schröder in Volume 3 of his *Vorlesungen* also developed quantificational logic, heavily influenced by Peirce, distinguishing first and higherorder quantificational logic. Geraldine Brady states that it was Schröder's attempt to code first-order statements into Peirce's relational connectives, which was the origin of Korselt's counter-example showing that this elimination of quantifiers in favor of relative product and the other relational operations does not always work.[40] Korselt produced a first-order statement not expressible in algebraic relational logic: the statement that the domain has at most four elements. Leopold Löwenheim started with Korselt's counter-example and compared the expressiveness

---

[40]See [**16**, p. 190].

of algebraic logic, first-order quantificational logic and higher order logic and proved that if a first-order statement has an infinite model, then it has a countable model [117]. The systems that Peirce and Schröder developed in the logic of relatives were used by him in his work, particularly in [116]. The tree of propositional valuations used by Löwenheim was used by both Jacques Herbrand and Kurt Gödel in their theses. So Peirce's algebra of relatives led through his quantificational logic to the notion of statements of first-order logic true in a domain. This treatment was expanded by Schröder and was the primary influence on Löwenheim's introduction of model-theoretic ideas into logic. Peirce also influenced logic in Poland and in particular the work of Mordchaj Wajsberg and Jan Lukasiewicz who looked at the axioms Peirce had formulated in his quantificational logic of 1885.[41]

Today's set theorists were also influenced by Peirce and Schröder, although it is essential to remember that the algebraic logicians were working in the Boolean part/whole theory of classes, not sets. Tarski and Givant in 1987 constructed an axiomatic system which presents set theory and number theory as sets of equations between predicates using the identity and 'element of' relations. This was inspired by the work of Peirce and Schröder, in particular Schröder's quest to express all elementary statements about relations as equations in the calculus of relations. Tarski was also familiar with Löwenheim's 1915 paper on the calculus of relations. Earlier in 1941 he had proved that the sentence expressing the existence of four elements is not expressible in the calculus of relations. This is directly relevant to Peirce's Reduction Thesis (i. e. that all polyadic relations can be reduced to monads, dyads and triadic relations) as it seems to provide a counterexample. However it must be realized that the counterexamples provided by Korselt, Löwenheim, Tarski and others of irreducible four-variable statements do not necessarily fully invalidate Peirce's Reduction Thesis, since they were primarily interested in the expressive power of equations and not with the irreducibility of relations. Tarski has shown only that equations concerning polyadic relations are not translatable into equations of three-variable fragments of first-order logic, not that polyadic relations are irreducible. It seems likely that Tarski was primarily influenced by Peirce here, as Schröder was concerned almost exclusively with dyadic relations.[42]

Some researchers such as Birkhoff [10, p. 9], Crapo and Roberts [29], and Salii [180, pp. 36–39] assert that lattices can be found in Peirce's work, in particular in

---

[41]See [94].
[42]See [2, p. 303].

[**153**]. These assertions center around the claim that Peirce held all lattices to be distributive, and invoke his correspondence with Huntington and with Schröder. But a close study of the sources shows that while all of the necessary pieces of apparatus were indeed present in [**153**]), they were not yet consolidated and formulated into a unified and coherent conception of lattices, as in [**184**] where he called this concept a *Dualgruppe*. Anellis and Houser [**3**, p. 5] claim that it is only after Huntington defined Boolean algebras, including the algebra of Peirce [**153**] as a complete complemented lattice as a complete complemented lattice, making it explicit for him, that Peirce fully recognized the lattice as a distinct mathematical entity.

The Russian logician P. S. Poretskii, also working in the algebraic tradition and a keen student of Boole, Jevons, and Schröder, in [**165**] declared that his papers represented the first attempt at constructing a complete and finished theory of qualitative argumentation and 'a completely original work' which allows for a transition from syllogisms to premises and for the possibility of solutions to problems within the theory of equations. Anellis and Houser claim that what Poretskii had in mind was the creation of a theory of quantification, which Peirce had begun to develop when he defined the existential and universal quantifiers in terms repectively of logical sums and products, in [**157**] a year before Poretskii. Styazhkin [**194**] also traces how P. S. Poretskii generalised and extended the logic of Boole, Jevons and Schröder. Schröder's method was unsatisfactory because it did not completely characterise the class of all inferences which can be drawn from a given logical equation unlike Poretskii's table of consequents. The fundamental concept of Poretskii's system was to 'solve' a logical equation by deducing from it all or some of its logical consequents. The complete solution of a given equation for Poretskii is that system of consequents which is equivalent to the equation itself. Developments are also occurring of Peirce's existential graphs.

Peirce's Reduction Thesis has proved fruitful for modern logicians. In particular, Burch [**22**] sought to prove this thesis in terms of existential graphs. Jacqueline Brunning's article "Triads and Teridentity" [**20**] uses the existential graphs to show the necessity for a third category of triads. However an algebraic proof has yet to be found. Another example is John Sowa's work [**191**] on conceptual graphs which are an extension of Peirce's existential graphs designed to accommodate linguistically relevant features. This arose out of logical considerations of semantic networks as applied to artificial intelligence. However in conceptual graphs, the point of quantification is shifted from the lines to the boxes. Conceptual graphs may prove to be the tool for analyzing and representing the many unsolved linguistic problems in representing plurals in linguistic theory.

Consideration of Peirce's influences on later logicians are of vital importance when we consider the historical fact that our modern notation for predicate logic came from Peirce's work through Schröder and Peano and not from Frege whose work was read only much later and whose notational system was never used by anyone else. Hilary Putnam also claims that Whitehead had come to his knowledge of quantification theory through Peirce.[43] However Russell wrote to Jourdain on 15 April 1910:[44]

> I read Schröder on Relations in September 1900 and found his methods hopeless.

Russell varied enormously in his position regarding extensional logic and intensional logic, but when defining a relation he emphasised the intensionalist view. Thus he did not like Peirce and Schröder's or even Peano's idea of a relation just as an ordered couple. Russell proposed 'the following [for the relation R on] classes: The class of terms which have the relation R to some term or other, which I call the class of referents with respect to R; and the class of terms to which some term has the relation R, which I call the class of relata.'[45] 'The intensional view of relations here advocated,' Russell states in *Principia Mathematica*, 'leads to the result that two relations may have the same extension without being identical.'[46] He also criticized Peirce:[47]

> In addition to the defects of the old Symbolic Logic, their method suffers technically ... from the fact that they regard a relation essentially as a class of couples, thus requiring elaborate formulae of summation for dealing with single relations ... this has led to a desire to treat relations as a kind of classes.

It should also be remembered that Russell was here refering to sets rather than the classes used by Peirce and Schröder.

---

[43]See [166].
[44]See [66].
[45]See [175, pp. 24].
[46]See [175, p. 24].
[47]*Ibid.*

An interesting question to consider at this point, is what influence Peirce had on the group of American mathematicians who were working in foundational studies during the period 1900-1930. These were labeled the 'American postulate theorists' by John Corcoran and identified by Michael Scanlan to include E. H. Moore, R. L. Moore, C. H. Langford, H. M. Sheffer, C. J. Keyser, E. V. Huntington, and O. Veblen. Scanlan [182] exemplifies their standards for axiomatizations of mathematical theories and their investigations of such axiomatizations with respect to metatheoretic properties such as independence, completeness, and consistency. The work of this school is important because the notion of categoricity of theories was first formulated by them as recognized by Tarski [196], and their work is also pertinent to the subsequent development of model theory. Scanlan does not trace any influences on this group other than that of Hilbert and Peano and merely comments 'the standards of work and approaches used in these papers seem essentially to have been developed within the American community of research mathematicians'. Huntington was well aware of Peirce's work. In particular, the proof of the distributivity law was given in [99], who reproduced it from a letter of Peirce's sent to him on 24 December 1903. As for Veblen, the director of his PhD. dissertation at the University of Chicago was E. H. Moore, who was not only editor of *Transactions of the American Mathematical Society*, but also a correspondent of Peirce.[48]

Another strand of Peirce's influence has thrown up a connection with Russell. The philosopher Josiah Royce (1855-1916), was a former student of Charles Peirce, and also Norbert Wiener's tutor at Harvard University. Wiener attended Royce's course on symbolic logic and later wrote a doctoral thesis comparing the logical systems of Schröder and Russell, with special reference to their treatment of relations. He went on to extend the logic of relations as used in *Principia Mathematica;* his best known contribution to logic being the definition of the ordered pair that reduced the logic of relations to the logic of classes. Another link with Royce concerns A. B. Kempe's study of the relation of geometry to logic [106]. In this work he sought to derive the order both of logical classes and of geometrical sets of points from assumptions in terms of a triadic relation $ac.b$ which may be read b IS BETWEEN a AND c.

Lewis states:[49]

─────────────

[48]See [98, p. 6].
[49]See [115, p. 365].

The triadic relation of Kempe is, then, a very powerful one, and capable of representing the most fundamental relations not only in logic but in all those departments of our systematic thinking where unsymmetrical transitive (serial) relations are important. In terms of these triads, Kempe states the properties of his 'base system', from whose order the relations of logic and geometry both are derived.

Royce mainly developed Kempe's 1886 theory and generalized his relations into polyadic relations which were then used to express the laws of the symbolic logic of classes [**174**]. So Kempe and Royce shared a general method which aimed to build sets of postulates which determine a class and the order of that class by selecting members from an initially generated and inclusive field.

7.6. The Divisions Between Algebraic and Mathematical Logic

The similarities and differences between the two traditions of algebraic and mathematical logic have been discussed in [**73**, p. 28–32]. The work of Peirce and Schröder was in general heavily eclipsed by Russell, Zermelo and others. Why did Peirce make no great effort to connect his work with that of Frege, Whitehead, Russell, Peano? He was certainly urged to do so by colleagues such as E. H. Moore and Christine Ladd-Franklin. The reason behind this could be that Peirce's sense of his own worth led him to consider that his logics were superior to anything developed by the mathematical logic camp. He also disagreed fundamentally with the logicist programme, so any idea of connection with their work in a systematic way might well have seemed pointless. Any interest shown by Peirce came at the end of his life, while the logicists largely dismissed his work as "so cumbrous and difficult that most of the applications which ought to be made are practically not feasible" [**175**, p. 24]. However, forty years later Russell accorded grudging praise as well as raising a major criticism commonly leveled against the algebraic logicians:[50]

> [Peirce] reminds one of a volcano spouting vast masses of rock, of which some, on examination, turn out to be nuggets of pure gold.

---

[50]See [**177**, p xvi].

Anellis and Houser[51] have pointed out that contemporary historians of logic have until recently, either ignored or downplayed the value of the algebraic logic tradition of the nineteenth century because it had been 'absorbed' into the more general mathematical logic of Whitehead and Russell's *Principia Mathematica*. Tarski himself in 1941 realized that this was at the cost of ignoring important mathematical content present in the algebraic tradition but not in the current mathematical logic. Many later logicians used the symbols for universal and existential quantification without reference to and seemingly unaware of the work of Peirce who had first introduced such notation. Mathematical logicians also strongly criticized the algebraic logicians for using the same symbol for class inclusion and implication. In this respect their following the set theoretical concepts of Cantor was an advantage to them.

Peirce himself took the view that the mathematical logic of Russell and Whitehead merely reformulated in a purely technical way, results in logic that he had already previously established. It was rather the antagonism of Russell and Frege that led to the separation of the algebraic tradition from that of mathematical logic. Whitehead and Peano were more sympathetic to the algebraic logicians and believed that many aspects of algebraic logic in particular that of quantification theory were useful.

P. Jourdain (1910–1913) and C. I. Lewis (1918) are among the few logicians and historians of logic working during the years when the *Principia* was first published, who continued to study the work of the algebraic logicians. Styazhkin (1964,1969) and his Soviet colleagues may have been influenced by the enormous contributions to the algebraic tradition by Poretskii, whose work Couturat (1914) called 'the improvement of the methods of Stanley Jevons and Venn.' However the current reassessment and further development of algebraic logic is continuing, mainly carried out by historians of Peirce as in [**98**].

The renewed interest in the work of Peirce (and in the development of algebraic logic) is encouraging when we consider that the culmination of mathematical logic in the *Principia Mathematica* has resulted in a further division in the branches of logic and mathematics, attributable in part to the disinterest of mathematicians in the *Principia* which contains a limited amount of mathematics, (e.g., the surprising omission of the calculus). The two communities of logicians and mathematicians largely lived apart, and 'each one thereby loses out from lack of mutual illumination.'[52] Logicians

---

[51]See [**3**].
[52]See [**68**, p. 79].

are now concerned primarily with philosophy and computing while philosophers of mathematics consider logic and set theory in isolation. This ignores the fruitful relationship between logic and mathematics successfully developed by the work of Charles Sanders Peirce.

ı

Bibliography

[1] H. Andréka, J. D. Monk, and I. Nemeti (eds.), Algebraic Logic, Proceedings of the conference in Budapest, 8-14 August, 1988, vol. 54, Amsterdam, Colloquia Mathematica Societatis Janos Bolyai, North Holland, 1991.

[2] Irving H. Anellis, Tarski's Development of Peirce's Logic of Relations, Studies in the Logic of Charles Sanders Peirce (Nathan Houser, Don D. Roberts, and James W. Van Evra, eds.), Indiana University Press, Indianapolis/Bloomington, 1997, pp. 271–303.

[3] Irving H. Anellis and Nathan Houser, Nineteenth Century Roots of Algebraic Logic and Universal Algebra, in Andréka et al. [1], pp. 1–36.

[4] Raymond Clare Archiblad (ed.), Benjamin Peirce, vol. 32, 1925.

[5] ———, Benjamin Peirce's Linear Associative Algebra and C. S. Peirce, Amer. Math. Monthly **34** (1927), 525–527.

[6] ———, Benjamin Peirce, Dictionary of American Biography **14** (1934), 1–20.

[7] James Mark Baldwin (ed.), Dictionary of Philosophy and Psychology, 1901, 2nd ed. in 3 vols.

[8] Richard Beatty, Peirce's Development of Quantifiers and of Predicate Logic, J. Formal Logic **10** (1969), 64–76.

[9] Eric Temple Bell, The Development of Mathematics, 2 ed., McGraw-Hill, New York, 1945.

[10] Garrett Birkhoff, Lattice Theory, Amer. Math. Soc., New York, 1940, 2nd Ed. 1948, 3rd Ed. 1979.

[11] George D. Birkhoff, Fifty Years of American Mathematics, Semicientennial Addresses of the American Mathematical Society, American Mathematical Society, New York, 1938, pp. 270–315.

[12] George Jordan Blazier, Marietta College Biographical Record of the Officers and Alumni, Marietta College, Marietta, Ohio, 1928.

[13] Innocentius M. Bochenski, A History of Formal Logic, Chelsea, New York, 1970, Translated and edited by Ivo Thomas.

[14] George Boole, The Mathematical Analysis of Logic, Being an Essay towards a Calculus of Deductive Reasoning, Macmillan, Cambridge, 1847, MAL.

[15] ———, An Investigation of the Laws of Thought, on which are founded the mathematical theories of Logic and Probabilities, Walton & Maberly, London, 1854, LA.

[16] Geraldine Brady, From the Algebra of Relations to the Logic of Quantifiers, in Houser et al. [**98**], pp. 173–192.

[17] Joseph Brent, Charles Sanders Peirce, A Life, Indiana University Press, Bloomington and Indianapolis, 1993.

[18] Chris Brink, On Peirce's Notation for the Logic of Relatives, Trans. Charles S. Peirce Soc. **14** (1978), 285–304.

[19] Jacqueline Brunning, Peirce's Development of the Algebra of Relations, Ph.D. thesis, University of Toronto, Toronto, 1980.

[20] ———, Genuine Triads and Teridentity, in Houser et al. [98], pp. 252–263.

[21] ———, Peirce on the Application of Relations to Relations, in Houser et al. [98], pp. 206–233.

[22] Robert W. Burch, A Peircean Reduction Thesis, Texas Tech University Press, Lubbock, 1991.

[23] Arthur W. Burks (ed.), Collected Papers of Charles Sanders Peirce: Reviews, Correspondence and Bibliography, vol. 8, Harvard University Press, Cambridge, 1958.

[24] Arthur W. Burks (ed.), Collected Papers of Charles Sanders Peirce: Science and Philosophy, vol. 7, Harvard University Press, Cambridge, 1958.

[25] William Elwood Byerly, Reminiscences, in Archiblad [4], pp. 5–7.

[26] Arthur Cayley, The Collected Mathematical Papers of Arthur Cayley, Cambridge University Press, Cambridge, 1896.

[27] Alanzo Church, Schröder's Anticipation of the Simple Theory of Types, Erkenntnis 9 (1939), 143–153.

[28] Glenn Clark, New Light on Peirce's Iconic Notation for the Sixteen Binary Connectives, in Houser et al. [98], pp. 304–33.

[29] Henry H. Crapo and Don D. Roberts, Peirce Algebras and the Distributivity Scandal, J. of Symbolic Logic 34 (1969), 153–154.

[30] M.J. Crowe, A History of Vector Analysis: The Evolution of the Idea of a Vectorial System, University of Notre Dame Press, Notre Dame, 1967.

[31] Joseph W. Dauben, Charles S. Peirce, Evolutionary Pragmatism and the History of Science, Centaurus 38 (1996), 22–82.

[32] Augustus De Morgan, First Notions of Logic, Preparatory to the Study of Geometry, Taylor & Walton, London, 1840.

[33] ———, Formal Logic: or The Calculus of Inference, Necessary and Probable, Taylor and Walton, London, 1847, Reprinted 1926 by A.E. Taylor, The Open Court Company, London.

[34] ———, On the Structure of the Syllogism, and on the Application of the Theory of Probabilities to Questions of Argument and Authority, Trans. Camb. Phil. Soc. 8 (1847), 379–408, Dated February 27, 1847. Read November 9, 1846. Reprinted in [xx], pp. 1–21.

[35] ———, On the Foundation of Algebra, No. IV., on Triple Algebra, Trans. Camb. Phil. Soc. 8 (1849), 241–253, Dated October 9, 1844. Read October 28, 1844.

[36] ———, On the Symbols of Logic, the Theory of the Syllogism, and in particular of the Copula, and the application of the Theory of Probabilities to some questions of evidence, Trans. Camb. Phil. Soc. 9 (1850), 79–127, Read February 25, 1850. Dated July 3, 1850. Reprinted in [xx], pp. 22–68.

[37] ———, On the Syllogism, No. III., and on Logic in General, Trans. Camb. Phil. Soc. 10 (1858), 173–230, Read February 8, 1858. Dated June 25, 1858. Reprinted in [xx], pp. 74–146.

[38] ———, Syllabus of a Proposed System of Logic, Walton & Maberly, London, 1860, Reprinted in [xx], pp. 147–207.

[39] ———, On the Syllogism, No. V., and on various points of the Onymatic System, Trans. Camb. Phil. Soc. 10 (1864), 428–487, Read May 4, 1863. Dated December 26, 1863. Reprinted in [xx], pp. 208–246.

[40] ———, On the Syllogism, and Other Logical Writings; ed. P. Heath, Routledge & Kegan Paul, London, 1966.

[41] Sophia Elizabeth De Morgan, Memoir of Augustus De Morgan by his wife Sophia Elizabeth De Morgan with Selections from his Letters, Longmans, London, 1882.

[42] Leonard Eugene Dickson, Definition of a Linear Associative Algebra by Independent Postulates, Trans. Amer. Math. Soc. **4** (1903), 1–27.

[43] Randall R. Dipert, Peirce's Propositional Logic, Review of Metaphysics **4** (1981), 569–595.

[44] _____, Book Review of *Studies In Logic*, Trans. Charles S. Peirce Soc. **20** (1984), 469–472.

[45] _____, Peirce, Frege, the Logic of Relations, and Church's Theorem, History and Philosophy of Logic **5** (1984), 49–66.

[46] _____, The Life and Work of Ernst Schröder, Modern Logic **1** (1991), 119–139.

[47] _____, The Life and Logical Contributions of O. H. Mitchell: Peirce's Gifted Student, Trans. Charles S. Peirce Soc. **30** (1994), no. 3, 515–542.

[48] _____, Peirce's Underestimated Place in the History of Logic, in Ketner [**108**], Plenary papers from Charles S. Peirce Sesquicentennial Congress, pp. 32–58.

[49] _____, Peirce's Philosophical Conception of Sets, in Houser et al. [**98**], pp. 53–76.

[50] Carolyn Eisele (ed.), The New Elements of Mathematics by C. S. Peirce, Mouton & Co. BV Publishers, The Hague, 1976, 4 vols. in 5 books.

[51] Charles William Eliot, Reminiscences of Peirce, in Archiblad [**4**], pp. 1–4.

[52] James Kern Feibleman, An Introduction to Peirce's Philosophy Interpreted as a System, Harper & Bros., New York, 1946.

[53] _____, An Introduction to Peirce's Philosophy, George Allen & Unwin Ltd., London, 1960.

[54] Della Fenster, The Development of the Concept of an Algebra: Leonard Eugene Dickson's Role, Rendiconti del Circolo Matematico di Palermo (1999), 59–122, Series II, Supplement 61.

[55] Max H. Fisch and Matthew E. Moore (eds.), Writings of C. S. Peirce: A Chronological Edition, vol. 1, Indiana University Press, Bloomington, August 1982, 1857–1866.

[56] Max H. Fisch and Matthew E. Moore (eds.), Writings of C. S. Peirce: A Chronological Edition, vol. 2, Indiana University Press, Bloomington, July 1984, 1867–1871.

[57] Max H. Fisch and Matthew E. Moore (eds.), Writings of C. S. Peirce: A Chronological Edition, vol. 3, Indiana University Press, Bloomington, April 1986, 1872–1878.

[58] Max H. Fisch and Matthew E. Moore (eds.), Writings of C. S. Peirce: A Chronological Edition, vol. 4, Indiana University Press, Bloomington, November 1989, 1879–1884.

[59] Max H. Fisch and Matthew E. Moore (eds.), Writings of C. S. Peirce: A Chronological Edition, vol. 5, Indiana University Press, Bloomington, December 1993, 1884–1886.

[60] Max H. Fisch and Matthew E. Moore (eds.), Writings of C. S. Peirce: A Chronological Edition, vol. 6, Indiana University Press, Bloomington, June 2000, 1886–1890.

[61] John Fiske, Edward Livingston Youmans, D. Appleton and Co., New York, 1894.

[62] Martin Gardner, Logic Machines and Diagrams, McGraw-Hill, New York, 1953.

[63] Carl Immanuel Gerhardt, Philosophischen Schriften von Leibniz, vol. 4, Weidmannsche Buchhardlung, Berlin, 1887.

[64] Kurt Gödel, Die Vollstandigkeit der Axiome des logischen Funktionenkalkuls, in Monatshefte fur Mathematik und Physik [**92**], 349–360, Trans. as 'The Completeness of the Axioms of the Functional Calculus of Logic' by S. Bauer-Mengelberg, in [textbf78], pp. 583–591.

[65] Ivor Grattan-Guinness, Wiener on the Logics of Russell and Schröder. An Account of his Doctoral Thesis, and of his Discussion of it with Russell, Annals of Science **32** (1975), 103–132.

[66] _____, Dear Russell – Dear Jourdain. A Commentary on Russell's Logic, Based on His Correspondence with Philip Jourdain, Duckworth, London, 1977.

[67] Ivor Grattan-Guinness (ed.), From the Calculus to the Set Theory, 1630–1910, An Introductory History, Duckworth, London, 1980.

[68] _____, Living Together and Living Apart. On the Interactions between Mathematics and Logics from the French Revolution to the First World War, South African Journal of Philosophy **7** (1982), no. 2, 73–82.

[69] _____, Psychology in the Foundations of Logic and Mathematics: the Cases of Boole, Cantor and Brouwer, History and Philosophy of Logic **3** (1982), 33–53.

[70] _____, The Correspondence between George Boole and Stanley Jevons, 1863–1864, South African Journal of Philosophy **12** (1991), 15–35.

[71] _____, Essay Review: Beyond Categories: The Lives and Works of Charles Sanders Peirce', a review of Brent 1993, Annals of Science **51** (1994), 531–538.

[72] _____, Benjamin Peirce's Linear Associative Algebra (1870): New Light on its Preparation and Publication, Annals of Science **54** (1997), 597–606.

[73] _____, Peirce between Logic and Mathematics, in Houser et al. [**98**], pp. 23–42.

[74] _____, The Search for Mathematical Roots, 1870–1940, Logics, Set theories and the Foundations of Mathematics from Cantor through Russell to Gödel, Princeton University Press, Princeton, 2000.

[75] Ivor Grattan-Guinness and Gérard Bornet (eds.), George Boole: Selected Manuscripts on Logic and its Philosophy, Birkhauser Verlag, Berlin, 1997.

[76] Judy Green, Christine Ladd-Franklin, in Grinstein and Campbell [**77**], pp. 121–128.

[77] Louise S. Grinstein and Paul J. Campbell (eds.), Women of Mathematics: A Biobibliographic Sourcebook, Greenwood Press, Connecticut, 1987.

[78] Theodore Hailperin, Boole's Logic and Probability, North-Holland, Amsterdam, 1976.

[79] William Hamilton (ed.), The Works of Thomas Reid, D. D., Now Fully Collected, with Selections from His Unpublished Letters. Preface, Notes and Supplementary Dissertations, by Sir W. Hamilton, Bart., [...] Prefixed, Stewart's Account of the Life and Writings of Reid; with Notes by the Editor, 1 ed., MachLachlan and Stewart & Co., Edinburgh, 1846, 1 vol. ending abruptly at page 914.

[80] William Hamilton (ed.), The Works of Thomas Reid, D. D., Now Fully Collected, with Selections from His Unpublished Letters. Preface, Notes and Supplementary Dissertations, by Sir W. Hamilton, Bart., [...] Prefixed, Stewart's Account of the Life and Writings of Reid; with Notes by the Editor, 6 ed., MachLachlan and Stewart & Co., Edinburgh, 1863, From this edition, in 2 vols., H. L. Mansel completes Hamilton's Dissertations and adds the Memoranda for Preface and the indexes.

[81] William Rowan Hamilton, On Quaternions, Proceedings Royal Irish Academy **3** (1847), 1–16.

[82] Charles Hartshorne and Paul Weiss (eds.), Collected Papers of Charles Sanders Peirce: Principles of Philosophy, vol. 1, Harvard University Press, Cambridge, 1931.

[83] Charles Hartshorne and Paul Weiss (eds.), Collected Papers of Charles Sanders Peirce: Elements of Logic, vol. 2, Harvard University Press, Cambridge, 1932.

[84] Charles Hartshorne and Paul Weiss (eds.), Collected Papers of Charles Sanders Peirce: Exact Logic, vol. 3, Harvard University Press, Cambridge, 1933.

[85] Charles Hartshorne and Paul Weiss (eds.), Collected Papers of Charles Sanders Peirce: The Simplest Mathematics, vol. 4, Harvard University Press, Cambridge, 1933.

[86] Charles Hartshorne and Paul Weiss (eds.), Collected Papers of Charles Sanders Peirce: Pragmatism and Pragmaticism, vol. 5, Harvard University Press, Cambridge, 1935.

[87] Charles Hartshorne and Paul Weiss (eds.), Collected Papers of Charles Sanders Peirce: Scientific Metaphysics, vol. 6, Harvard University Press, Cambridge, 1935.

[88] Hebert E. Hawkes, Estimate of Peirce's Linear Associative Algebra, Amer. J. Math. **24** (1902), 87–95.

[89] _____, Hypercomplex Number Systems in Seven Units, Amer. J. Math. **26** (1904), 223–242.

[90] Peter Heath (ed.), On the Syllogism, and Other Logical Writings, London, Routledge & Kegan Paul, 1966.

[91] Peter Heath, On the Syllogism, and Other Logical Writings, [**90**], pp. i–xix.

[92] Jean van Heijenoort (ed.), From Frege to Gödel: A Source Book in Mathematical Logic, 1879–1981, Harvard University Press, Cambridge, 1967.

[93] Jacques Herbrand, Recherches Sur la Theorie de la Demonstration, Ph.D. thesis, Univ. Paris, Paris, 1930, Chapter 5 trans. by B. Dreben and J. van Heijenoort as "Investigations in Proof Theory: The Properties of True Propositions" in Heijenoort:1967, pp. 529–581.

[94] Henry Hiz, Peirce's Influence on Logic in Poland, in Houser et al. [**98**], pp. 264–270.

[95] Christopher Hookway, Benjamin Peirce, Routledge & Kegan Paul, London, 1985.

[96] Nathan Houser, Peirce's Early Work on the Algebra of Logic: Remarks on Zeman's Account, Trans. Charles S. Peirce Soc. **23** (1987), 425–440.

[97] _____, The Schröder-Peirce Correspondence, Modern Logic **1** (1990), 206–236.

[98] Nathan Houser, Don D. Roberts, and James van Evra (eds.), Studies in the Logic of Charles Sanders Peirce, Bloomington and Indianapolis, Indiana University, 1997.

[99] Edward V. Huntington, Sets of Independent Postulates for the Algebra of Logic, Trans. Amer. Math. Soc. **5** (1904), 288–309.

[100] Edward V. Huntington and Christine Ladd-Franklin, Symbolic Logic, The Americana. A Universal Reference Library, vol. 9, The Americana Co., 1905, Reprinted in 1934 Ed., vol 17, pp. 568–573.

[101] Dorothea Jameson Hurvich, Christine Ladd-Franklin, Notable American Women, vol. 2, Harvard University Press, Cambridge, 1971, pp. 354–356.

[102] Alan J. Iliff, The Matrix Representation in the Development of the Quantifiers, in Houser et al. [**98**], pp. 193–205.

[103] William Stanley Jevons, Pure Logic, or the Logic of Quality Apart from Quantity with Remarks on Boole's System and on the Relation of Logic and Mathematics,, Stanford, London, 1864.

[104] Hans Kamp, Events, Discourse Representations, and Temporal References, Languages **64** (1891), 39–64.

[105] Sir Alfred Bray Kempe, A memoir on the theory of mathematical form, Phil. Trans. Roy. Soc. London (A) **177** (1886), 1–70.

[106] _____, On the Relation between the Logical Theory of Classes and the Geometrical Theory of Points, Proc. Lond. Math. Soc. **21** (1890), 147–182.

[107] Angus Kerr-Lawson, Truth and Idiomatic Truth in Santayana, Transactions of the Charles S. Peirce Society **33** (1997), no. 1, 91 – 111.

[108] Kenneth Laine Ketner (ed.), Peirce and Contemporary Thought, Fordham University Press,, New York, 1995, Plenary papers from Charles S. Peirce Sesquicentennial Congress.

[109] William Kneale and Martha Kneale, The Development of Logic, Oxford University Press, Oxford, 1962.

[110] Christine Ladd-Franklin, On the Algebra of Logic, in Marquand et al. [**120**], Introduction by C. S. Peirce, pp. 17–71.

[111] _____, Review of Schröder 1890, Mind **1** (1892), 126–132.

[112] Johann Heinrich Lambert, Logische und philosophische Abhandlungen; ed. Johann III Bernoulli, Selbstverlag, Berlin, 1782 & 1787, 2 volumes.

[113] Victor F. Lenzen, The Contributions of Charles S. Peirce to Linear Algebra, in Riepe [**170**], pp. 239–254.

[114] _____, Peirce's Theoremic/Corollarial Distinction and the Interconnections between Mathematics and Logic, in Houser et al. [98], pp. 85–110.

[115] Clarence Irving Lewis, A Survey of Symbolic Logic, Univ. California Press, Berkeley, 1918.

[116] L. Löwenheim, Einkleidung der Mathematik im SchröderschenR elativkalkul, J. of Symbolic Logic 5 (1940), 1–15.

[117] Leopold Löwenheim, Über Moglichkeiten im Relativkalkü 1',, Mathematische Annalen 76 (1915), 447–470, Trans. S. Bauer-Mengelberg as "On Possibilities in the Calculus of Relatives", in van Heijenoort 1967, pp. 228–251.

[118] Hugh MacColl, Calculus of Equivalent Statements, Proc. Lond. Math. Soc. 9 (1877), 77–186, Series 1.

[119] Dumas Malone (ed.), Dictionary of American Biography, vol. xiv, Scribner Press, New York, 1934.

[120] Allan Marquand, Christine Ladd-Franklin, Oscar Howard Mitchell, and Benjamin Ives Gilman (eds.), Studies in Logic, By Members of the Johns Hopkins University, Foundations of Semiotics, Little, Brown, Boston, 1883, Introduction by C. S. Peirce.

[121] Richard Milton Martin, Peirce's Logic of Relations and Other Studies, The Peter De Ridder Press, Lisse, The Netherlands, 1979.

[122] Daniel D. Merrill, De Morgan, Peirce, and the Logic of Relations, Trans. Charles S. Peirce Soc. 14 (1978), 247–284.

[123] _____, The 1879 Logic of Relatives Memoir, W2, xlii–xlviii, 1986.

[124] _____, Augustus De Morgan and the Logic of Relations, Kluwer Academic Publishers, Dordrecht, The Netherlands, 1990.

[125] _____, Relations and Quantification in Peirce's Logic, 1870–1885, in Houser et al. [98], pp. 158–172.

[126] Emily Michael, Peirce's Early Study of the Logic of Relations, 1865–1867, Trans. Charles S. Peirce Soc. 10 (1974), 63–67.

[127] _____, An Examination of the Influence of Boole's Algebra on Peirce's Developments in Logic, Nodre Dame J. Formal Logic 20 (1979), 801–806.

[128] Oscar Howard Mitchell, On a New Algebra of Logic, in Marquand et al. [120], Introduction by C. S. Peirce, pp. 72–103.

[129] Murray G. Murphey, The Development of Peirce's Philosophy, Harvard University Press, Cambridge, 1961, Reprinted 1993 Hackett: Philadelphia.

[130] Hubert Anson Newton, Benjamin Peirce, Amer. J. Science 22 (1881), 167–178.

[131] Lubos Nový, Origins of Modern Algebra, Noordhoff Internation, Leyden, The Netherlands, 1973, Trans. J. Tauer.

[132] _____, Benjamin Peirce's Concept of Linear Algebra, Acta historiae naturalium necnon techicarum (1974), 211–231, Special Issue 4, Prague.

[133] Maria Panteki, Relationships between Algebra, Differential Equations, and Logics in England, 1800–1860, Phd., CNAA London, London, 1992.

[134] Karen Hunger Parshall (ed.), James Joseph Sylvester: Life and Work in Letters, Oxford University Press, Oxford, 1998.

[135] George Peacock, Treatise of Algebra, J. and J. J. Deighton,, Cambridge, 1830.

[136] Volker Peckhaus, Ernst Schröder und die "Pasigraphischen Systeme" von Peano und Peirce, Modern Logic (1991), 174–205.

[137] Benjamin Peirce, Linear Associative Algebra, Private lithograph ed., Washington D. C., Ed. C.S. Peirce, 1870.

[138] _____, On the Uses and Transformations of Linear Algebra, Proceedings of the American Academy of Arts and Sciences **2** (1875), 397–400.

[139] _____, Addendum: On the Uses and Transformations of Linear Algebra, Amer. J. Math. **4** (1881), 216–221.

[140] _____, Linear Associative Algebra, Amer. J. Math. **4** (1881), 97–215.

[141] Charles Sanders Peirce, The Charles S. Peirce Papers, Manuscript collection in the Houghton Library, Harvard University, 1849–1914.

[142] _____, Harvard Lecture III: The Logic of Relatives, MS 96, A1, pp. 189–199, 1865.

[143] _____, Harvard Lecture VI: Boole's Calculus of Logic, MS 100, A1, pp. 223–239, 1865.

[144] _____, An Improvement in Boole's Calculus of Logic, Proceedings of the American Academy of Arts and Sciences **7** (1867), 250–261, Reprinted CP 3.1–3.19 and W2,12–23.

[145] _____, On the Natural Classification of Arguments, Proceedings of the American Academy of Arts and Sciences **7** (1867), 261–287, Presented April 9, 1867. Reprinted (CP 2.461516), (W 2:2349).

[146] _____, Upon the Logic of Mathematics, Proceedings of the American Academy of Arts and Sciences **7** (1867), 402–412, Repr. CP 3.20–3.44 and W2,59–69.

[147] _____, Note 4, 1868, MS 152. Reprinted in W2 (only), 1984, 88–92.

[148] _____, Upon the Logic of Mathematics, Proceedings of the American Academy of Arts and Sciences **7** (1868), 287–298, Reprinted CP 1.545–1.559 and W2,49–59.

[149] _____, Description of a Notation for the Logic of Relatives, Resulting from an Amplification of the Conceptions of Boole's Calculus of Logic, Memoirs of the American Academy of Arts and Sciences **9** (1870), 317–378, Reprinted Weiss 3.45–3.149 and Fisch 359–429.

[150] _____, On the Application of Logical Analysis to Multiple Algebra, Proceedings of the American Academy of Arts and Sciences **10** (1875), 392–394, Reprinted CP 3.15–3.151 and in W3, 177–179.

[151] _____, On the Algebraic Principles of Formal Logic, 1879, MS 348. Reprinted in W4, 21–37.

[152] _____, A Boolian Algebra with One Constant, 1880, MS 378. Reprinted (only) in W4,218–221.

[153] _____, On the Algebra of Logic, Amer. J. Math. **3** (1880), 392–394, Reprinted in CP 3.154–251 and in W4, 163–209.

[154] _____, On the Logic of Number, Amer. J. Math. **4** (1881), 85–95, Reprinted in CP 3.252–288 and in W4, 299–309.

[155] _____, Brief Description of the Algebra of Relatives, 1882, Privately printed brochure, Baltimore Reprinted in CP 3.306–322 and in W4,328–333.

[156] _____, The Chemical Theory of Interpenetration, American Journal of Science and Arts **35** (1883), 78–82.

[157] _____, The Logic of Relatives (Note B), 1883, Reprinted in CP 3.328–3.358 and in W4, 453–466.

[158] _____, On the Algebra of Logic, Amer. J. Math. **7** (1885), 80–202, Reprinted in CP 3.359–3.403 and in W4, 162–190.

[159] _____, The Regenerated Logic, The Monist **7** (1896), 19–40, Reprinted in CP 3.425–3.455.

[160] _____, Logic of Relatives, The Monist **7** (1897), 161–217, Reprinted in CP 3.456–3.552.

[161] _____, Logic, (exact), in Baldwin [**7**], Reprinted in part in CP 3.616–3.625.

[162] _____, Nomenclature and Divisions of Dyadic Relations, Lowell Lectures, 1903, Reprinted in CP 3.571–3.608.

[163] _____, Notes on Symbolic Logic and Mathematics, in Baldwin [**7**], Vol. 1, p. 518.

[164] An Phan, Christine Ladd-Franklin, Biographies of Women Mathematicians, 1995.

[165] P. S. Poretskii, On methods of solving logical equalities and on an inverse method of mathematical logic, Collection of protocols of sessions of the Fiz-Mat section of the Society of Natural Scientists at Kazan Univ., vol. 2, Kazan University, 1884, pp. 161–330.

[166] Hilary Putnam, Peirce the Logician, Historia Mathematica **9** (1982), 290–301.

[167] Helen M. Pycior, Benjamin Peirce's Linear Associative Algebra, Isis **70** (1979), 537–551.

[168] ———, British Synthetic vs. French Analytic Styles of Algebra in the Early American Republic, in Rowe and McCleary [**173**], pp. 125–154.

[169] Rush Rhees, Introduction, in Heath [**90**], pp. 1–10.

[170] D. M. Riepe (ed.), Phenomenology and Natural Existence: Essays in Honor of Marvin Farber, State Univ. New York Press, Albany, 1973.

[171] Don D. Roberts, The Existential Graphs of Charles S. Peirce, Mouton, The Hague, 1973.

[172] Richard S. Robin, Annotated Catalogue of the Papers of Charles S. Peirce,, Univ. Mass. Press, Amherst, USA, 1967.

[173] David Rowe and John McCleary (eds.), The History of Modern Mathematics, vol. 2, Academic Press, San Diego, 1989.

[174] Josiah Royce, The Relation of the Principles of Logic to the Foundations of Geometry, Trans. Amer. Math. Soc. **6** (1905), 353–415.

[175] Bertrand Russell, The Principles of Mathematics, Cambridge University Press, Cambridge, 1903.

[176] ———, The Principles of Mathematics, 2 ed., Allen & Unwin, London, 1937.

[177] ———, Introduction, [**52**], pp. xv–xvi.

[178] Bertrand Russell and Alfred North Whitehead, Principia Mathematica, Cambridge University Press, Cambridge, 1910, 3 vols.

[179] ———, Principia Mathematica, 2 ed., Cambridge University Press, Cambridge, 1925–1927.

[180] V. N. Salii, Lattices with Unique Complements, Amer. Math. Soc., 1988, Translated by G.A. Kandall.

[181] John Turner Sargent (ed.), Sketches and Reminiscences of the Radical Club, J. R. Osgood & Co., Boston, 1880.

[182] Michael Scanlan, Who Were The American Postulate Theorists?, Journal of Symbolic Logic **46** (1991), 981–1002.

[183] Ernst Schröder, Der Operationskreis des Logikkalkuls, Teubner, Leipzig., 1877.

[184] ———, Vorlesungen über die Algebra der Logik, vol. 1, Teubner, Leipzig, 1890.

[185] ———, Vorlesungen über die Algebra der Logik, vol. 2, Teubner, Leipzig, 1891.

[186] ———, Vorlesungen über die Algebra der Logik, vol. 3, Teubner, Leipzig, 1895.

[187] ———, Vorlesungen über die Algebra der Logik, vol. 4, Teubner, Leipzig, 1905.

[188] ———, Vorlesungen über die Algebra der Logik, vol. 3, Chelsea, New York, 1966.

[189] James Byrnie Shaw, Synopsis of Linear Associative Algebra, Carnegie Institution of Washington, Washington D. C., 1907.

[190] Arthur Thomas Shearman, The Development of Symbolic Logic, Williams & Norgate, London, 1906.

[191] John F. Sowa, Matching Logical Structure to Linguistic Structure, in Houser et al. [**98**], pp. 418–444.

[192] William Spottiswoode, Remarks on Some Recent Generalizations of Algebra, Proc. Lond. Math. Soc. **4** (1872), 147–164.

[193] Leslie Stephen and Sidney Lee (eds.), The Dictionary of National Biography, vol. 5, Oxford University Press, Oxford, 1921.

[194] Nicholai Ivanovich Styazhkin, History of Mathematical Logic from Leibniz to Peano, M. I. T. Press, Cambridge, 1969.

[195] Henry Taber, On Hypercomplex Number Systems, Trans. Amer. Math. Soc. **5** (1904), 509–548.

[196] Alfred Tarski, Sur les ensembles finis, Fundementa Mathematicae **6** (1924), 45–95.

[197] Alfred Tarski and Steven Givant, A Formalization of Set Theory without Variables, Amer. Math. Soc., Providence, 1987.

[198] I. Thomas, Boole's Concept of Science, Proceedings of the Royal Irish Academy **57** (1955), no. 6, 88–96.

[199] James van Evra, Logic and mathematics in Charles S. Peirces "Description of a notation for the logic of relatives", in Houser et al. [**98**], pp. 147–157.

[200] John Venn, Boole's Logical System, Mind **1** (1876), 479–491.

[201] Richard Whatley, Elements of Logic, Comprising the substance of the Article in the Encyclopaedia Metropolitana., J. Mawman, London, Repr. 1975, New York. Repr. 1988, with introduction by P. Dessi, Bologna.

[202] Shea Zellweger, Semeiotics, Good Symmetry and the Logic of Propositions, Symmetry, Culture & Science **3** (1992), 76–77.

[203] ———, Studies in the Logic of Charles Sanders Peirce, in Houser et al. [**98**], pp. 334–386.

[204] J. Jay Zeman, Peirce's Graphs – the Continuity Interpretation, Trans. Charles S. Peirce Soc. **4** (1968), 144–154.

[205] ———, Peirce's Logical Graphs, Semiotic **12** (1974), 239–256.

[206] ———, Peirce's Philosophy of Logic, Trans. Charles S. Peirce Soc. **22** (1986), 1–22.

Made in the USA
San Bernardino, CA
27 April 2015